T0003227

ISRAEL 201

YOUR NEXT-LEVEL GUIDE TO THE MAGIC, MYSTERY, AND CHAOS! OF LIFE IN THE HOLY LAND

JOEL CHASNOFF & BENJI LOVITT

gefen גפן
publishing house בית הוצאה לאור
JERUSALEM • NEW YORK   Est. 1981

Copyright © Joel Chasnoff and Benji Lovitt

Jerusalem 2024/5784

All rights reserved. No part of this publication may be translated,
reproduced, stored in a retrieval system or transmitted, in any form
or by any means, electronic, mechanical, photocopying, recording or
otherwise, without express written permission from the publishers.

Cover Design: Zoe Norvell

Set in Arno Pro by Raphaël Freeman MISTD, Renana Typesetting

ISBN: 978-965-7801-28-4

5 7 9 8 6

Gefen Publishing House Ltd.	Gefen Books
6 Hatzvi Street	c/o Baker & Taylor Publisher Services
Jerusalem 9438614	30 Amberwood Parkway
Israel	Ashland, Ohio 44805
972-2-538-0247	516-593-1234
orders@gefenpublishing.com	orders@gefenpubl. .ѣ ng.com

orders@gefenpublishing.com

www.gefenpublishing.com

Printed in Israel

Library of Congress Control Number: 2022921122

For Ruti Halpern, David Keren, and Simcha Assaf
Leibovich – the three people most responsible for my
connection to this magical (and so often maddening) country.

J.C.

⁊⊘

To my parents, Zelene and Robert, for always
supporting me and my decisions, even when every
move took me further away. I love our video calls.

B.L.

Contents

Chapter 2: Jewish Life in a Jewish State

Chapter 3: The Hebrew Language   74

Chapter 4: Government, Social Policy, and Education   91

Preface

DAVID BEN-GURION, ISRAEL'S FIRST PRIME MINISTER, famously said, "We know we'll be a normal country when Jewish prostitutes and Jewish thieves conduct their business in Hebrew."

We love this quote. Not just because Ben-Gurion says "prostitute," but because it so perfectly captures the contradiction inherent in the idea of a modern Jewish state: the desire, on the one hand, to be a nation like any other – with teachers and farmers, garbagemen and firemen, athletes and mayors – and, at the same time, a recognition that this country can't *help* but be special, by virtue of being the only one on earth where, for the first time in two thousand years, the Jewish people live in self-determination, proudly, as Jews.

Over the years, Jewish education has tended to focus on the "special," the cultural, the mythological Israel. The Israel that inspires us to plant trees in Israeli forests and attend Israel Day fairs where we eat falafel, dance the hora, and get our faces painted with blue and white flags. Deeper conversation was mostly about the biggest questions facing the country – first, regarding the greater Arab-Israeli conflict and, in the twenty-first century, the conflict with the Palestinians. In recent years, we've learned about Israel as an incubator of start-ups that brought us such groundbreaking innovations as drip irrigation, the cherry tomato, and Waze.

Don't get us wrong – existential questions are important, and bite-sized vegetables (or are they fruit?) should be celebrated. But to focus only on the successes, threats, and the "special" part of Ben-Gurion's vision is to miss out on the living, breathing, "normal" Israel. If we step back and reevaluate the other side of his quote: What about the metaphorical prostitutes and thieves?

In *Israel 201*, we pull back the curtain to unveil a more complete and comprehensive portrait of Israel that extends beyond the headlines and narratives we typically encounter. The Israel that wakes up in the morning to go to work or school (that is, when the teachers aren't on strike). The Israel that lacks enough publicly owned land to ensure affordable home ownership for all but the wealthiest members of society. The Israel that struggles to limit its stray cat population because of laws in the Torah related to *tza'ar ba'alei chayim*, the ethical treatment of animals.

Our goal is to provide you a deeper understanding of this magical and chaotic country, a country about which, despite taking up so much headspace in the minds of citizens of the world, little is known and even less understood.

Hence the title: *Israel 201* represents an advanced look at Israel, beyond the basics of the typical introductory "101" course. If you've ever wondered why Israelis go to synagogue in jeans, what games Israeli schoolkids play at recess, or how actually living in Israel is different from your standard two-week visit, this book is for you.

That said, you don't need to know much – or really anything – about Israel to benefit from this book. If you've ever asked, "Where do I begin?!" intimidated by the overwhelming number of resources, media, and news sites, *Israel 201* offers a breadth and depth into Israeli society at a level of detail you'll be hard-pressed to find anywhere else.

How This Book Works

Israel 201 contains eight chapters, broken down into around seventy-five topics.

In chapter 1, we explore the Israeli personality, for the simple fact that you can't truly understand *anything* about this country until you understand its people's psyche. Israelis' unique worldview is a product of their past and current circumstances; it colors everything from politics and government to show business, parenting, office culture, and even the angry text message you might receive after you tell your date that you're "just not feeling it."

From there, we venture into what we believe are the topics most critical to making sense of this place: Judaism's role in daily life; the Hebrew language; government, social policy, and education; the economy, work, and work-life balance; the Israel Defense Forces (IDF); the arts, culture, sports, and leisure; and aliyah, immigration to Israel.

In our afterword, we look into our crystal ball to ask the question that's on the minds of millions in Israel and around the world: What does the future hold for this tiny country?

We've included a glossary of Hebrew words you'll need to know. Please turn to the back of the book first to familiarize yourself with these important terms.

You'll also find at the back a Sabra Bingo game, for your enjoyment.

As for chapters that didn't make the cut? Beach culture, the Ethiopian-Jewish holiday Sigd, the country's unofficial hummus capital Abu Ghosh, the connection between an Oscar-winning movie producer and Dimona's nuclear reactor (we're not confirming its existence – actually, forget we mentioned it ...). While we enjoyed coming up with just some of the countless possibilities, we quickly realized that picking the "right" topics was an impossible task. So we narrowed it down to those we most wanted to research and didn't look back. (The upside: We have plenty left over for *Israel 301*.)

Who Are We, and Why Us?

Joel and Benji both grew up in the United States and made aliyah (immigrated to Israel) in their early twenties and thirties, respectively.

Benji's decision was the culmination of his Jewish upbringing in his hometown of Dallas, Texas, and involvement in the Young Judaea youth movement and summer camps – first as a camper, then as a summer staff member, and ultimately as a year-round professional, operating Israel summer programs for teens. Each visit to Israel became more meaningful than the last and gave him a feeling of belonging and joy that made his spirit come alive. In the summer of 2005, at the end of his annual Young Judaea work trip to Israel, his Israeli colleague Ofra pulled him aside and said, "Don't go. You belong here. You are one of us." Benji cried. One year later, he was Israeli. Since then, he has lived mostly in Tel Aviv, with a few years in Jerusalem in between. Before you ask which he prefers, know that his answer is, "Why do I have to choose?"

Joel made aliyah for similar reasons. Raised outside Chicago, Illinois, he grew up with a strong Jewish identity, including nine years at Solomon Schechter Day School and summers at Camp Ramah. But it was a six-week teen tour to Israel at age seventeen that changed him. Though he'd visited before, something about this trip made him feel at home. He vowed not only to return, but to do as Israelis do and serve in the army. In the summer of 1997, shortly after university, Joel made aliyah and was drafted into the Armored Corps. He served as a tank gunner in a platoon of eighteen-year-olds who'd graduated high school just weeks earlier, under officers three years younger than he was, a story that became the subject of Joel's comedic memoir *The 188th Crybaby Brigade*. After his military service, Joel and his Israeli wife Dorit returned to the US, having four kids along the way. In 2016, the entire family moved back and settled in Ra'anana.

For the past twenty years, both Joel and Benji have made their living as professional comedians. Knowing that, you might assume that this book is a nonstop romp of hilarity and jokes. Indeed, we do hope to make you laugh here and there as we share the absurdities of life in twenty-first-century Israel.

That said, this is not, first and foremost, a humor book. In *Israel 201*, we exchange our comedian hats for a pair of high-powered magnifying glasses as we inspect the country from every conceivable angle and do that other thing comedians love to do: *observe*. We tried to write the book in a way that's engaging and easy to digest, as if you're sitting around talking with a couple of good friends.

Having lived significant portions of our lives in both the US and Israel, we bring a unique perspective to this book. On one hand, we are insiders who have collectively participated in every major aspect of Israeli life: being hired and fired from jobs, navigating the health care system, scampering into bomb shelters in the middle of the night, opening and closing bank accounts, overpaying for cars, dating, and, in Joel's case, getting married and raising kids here, one of whom is currently an officer in the IDF.

At the same time, we are also outsiders, immigrants who've spent decades in the United States with the life experience to compare, contrast, and perceive differences between our two homes, both large and small.

In *Israel 201*, we draw from our dual identities to shed light on the many differences between life in Israel and abroad and to introduce you to the numerous aspects of Israel that most people, including many natives, don't know about.

That being said, as two self-deprecating American-born expats, we happily admit we don't know it all. For that reason, we interviewed dozens of native Israeli experts from all walks of life, from award-winning TV executives to world-class athletes, internationally recognized professors, military officers so high in rank they couldn't give their names, and everyday people of all ages, for their perspective on the country. In addition to knowledge and expertise, these personal interactions breathed an element of heart, soul, and warmth into the narrative that we neither expected nor imagined upon embarking on this journey.

Sorry in Advance

We can pretty much guarantee that by the time this book reaches your hands, some things about Israel will have changed, which means that some of what's in this book could already be out of date – and for that, we apologize. Just as we were finalizing the original manuscript, the country's government collapsed (again), necessitating changes to the chapters on prime ministers and elections. By now, we may have voted several more times (please, God, no).

Likewise, our chapter on "hot-button issues" needed an emergency revision when, without warning, Israel changed its laws to make abortion more accessible and less burdensome, the week after the US Supreme Court overturned Roe v. Wade. Changes to laws related to conversion, marriage, and other life-cycle events are being debated as we speak. For all we know, by the time you read this, Israel's prime minister could be hugging Iran's supreme leader on the White House lawn. As for the name of that prime minister, your guess is as good as ours.

And, of course, 2023 brought the most tempestuous year in Israel's history, first with the judicial reform crisis, then the October 7th massacre – the worst day in Israel's seventy-five years – followed by the resultant war, which continues as this goes to print. This edition of *Israel 201* includes a new afterword with Professor Noah Efron about how these events may (or may not) have changed Israel forever. Content within specific sections of the book has also been updated to reflect the new reality of post–October 7th Israel.

Israelis, Non-Israelis, and Anglos

Much of this book explores the Israeli personality and how Israelis behave differently from others.

And for this, we present a disclaimer: it's impossible to discuss any people without generalizing. All peoples are complicated, their personalities a mix of values, behaviors, and experiences resulting from their birthplaces and the lands they pass through.

For that matter, even the word *Israelis* is fraught. The term encompasses Ashkenazi Jews with roots in Eastern Europe, Mizrachi Jews from countries such as Morocco and Yemen, Ethiopians, Russians, immigrants from myriad other countries, and, of course, the more than 20 percent of Israel's population who aren't Jewish.

Still, there's no denying that national characters exist. Just as it's fair to say that Americans as a whole are different from the French or the Dutch, so, too, is it true that Israelis possess a set of values and behavior unique to them. And even though Jews and Arabs/Palestinians in Israel have more in common than you might imagine, when we use the term *Israelis* in this book, we are, more often than not, referring to Jewish Israelis.

There's also a question of frame of reference: you can't draw meaningful conclusions about Israelis without having a population to compare them to. But who exactly is that?

Most likely, our readers are Americans like us, the biggest Jewish community outside of Israel. That said, we've lived abroad long enough to realize (and be reminded) that – speaking of national characteristics – Americans can be known to ignore the rest of the world (guilty as charged), so we tried to keep *all* of you in mind when writing. In this book, we often use *Westerners* or the popular Israeli term *Anglos* as a catch-all for natives of English-speaking countries such as the United States, Canada, the United Kingdom, South Africa, and Australia.

The O-Word

As you may notice, the word *occupation*, in the context of Israel, appears only once in this entire book aside from this sentence. (It arose in our interview with Avi Issacharoff, cocreator of the TV show *Fauda*, which, if you've seen the show, will not be surprising.)

Some of you may find this shocking or problematic; indeed, for many around the world, Israel *is* the conflict with the Palestinians.

But that's one of the reasons we don't address it in *Israel 201*. As impactful, significant, and tragic as the conflict is, Israel is so much more than that.

Secondly, the topic is simply beyond the scope of this book; any attempt to address its messy complexity within these pages would have been shallow and inappropriate.

That leads us to still another reason: many fantastic books about the

Palestinian-Israeli conflict already exist, and if someone can improve upon them, it's not us. One of our greatest ambitions for this book is that it will inspire readers to explore Israel more deeply; to that end, those who want to know more can choose from a number of excellent works. Just a few include Yossi Klein Halevi's *Letters to My Palestinian Neighbor*, Noa Tishby's *Israel: A Simple Guide to the Most Misunderstood Country on Earth*, Daniel Gordis's *Israel: A Concise History of a Nation Reborn*, Ari Shavit's *My Promised Land: The Triumph and Tragedy of Israel*, and Daniel Sokatch's *Can We Talk about Israel? A Guide for the Curious, Confused, and Conflicted*, each with its own particular point of view, all along the political spectrum.

To be clear: we are not minimizing the gravity of Israel's conflict with the Palestinians, nor assessing the level of blame Israel shoulders. We are saying that it's complicated, it warrants acknowledgment...and it's not what *this* book is about.

B'hatzlachah! (Good Luck!)

One of the tour guides we interview in this book aims to send tourists home with more questions than answers.

In many ways, that's our goal too. Or at the very least, to answer some of your current questions.

"Why are Israelis so blunt?"

"What's it like going to the army?"

"How is Israeli comedy different from traditional Jewish humor?"

Our ultimate hope is that you will feel a new, stronger connection to the country. Our own relationships with Israel, like those of all *olim* (immigrants to Israel), have included their fair share of ups and downs. Writing this book has rekindled our love for the country in a way we didn't anticipate, in large part thanks to the truly incredible people we met along the way. We're talking about the people who make this country more than a concept, idea, or dream to be debated, discussed, celebrated, and criticized, but rather, a living, breathing, *actual* place, that manages to somehow be both normal and special at the same time:

the think tank researchers committed to improving governance and the well-being of all citizens; the educators and teachers who go above and beyond to raise the next generation of leaders, despite wages that come nowhere close to justifying their efforts; the nonprofit workers and volunteers who help at-risk populations...and, yes, even the prostitutes who do their business in Hebrew. (Some of it, anyway.)

Ben-Gurion would be so proud.

Pre-Course Quiz
How Israeli Are *You*?

WE'LL BEGIN OUR DEEP DIVE INTO THE ISRAELI PSYCHE IN A moment. But first, let's find out: How Israeli are *you*?

Choose one answer per question and keep track for scoring purposes at the end.

The Israeliness Quiz

1. You're stopped at a red light. The light turns green, but the car in front of you doesn't move. You:
   a. Wait patiently for the car to go.
   b. Beep your horn lightly.
   c. Blare your horn loudly and continuously, shouting "*Nu!*" until the moron finally goes.

2. You're a father attending your thirteen-year-old son's bar mitzvah. You wear:
   a. Your best suit and tie.
   b. Sport coat, slacks, and tie.
   c. Slacks and button-down shirt, halfway open to reveal as much chest hair as possible.

3. Your cousin is getting married. As a gift, you give:
   a. The china set from her FOX Home wedding registry.
   b. A gift certificate to FOX Home.
   c. A thousand shekels cash in a blank envelope with your personal message scribbled illegibly on the outside (why waste money on a pretentious, impersonal greeting card?).

4. You see a LinkedIn post for a job you're dying to get. After reading the full description, you:
   a. Decide not to apply because you fear you're not qualified.
   b. Send a message saying you'd love an opportunity to interview.
   c. Send a very confident message that you're extremely qualified and they'd be nuts not to hire you on the spot.

5. You're driving on a side street when a traffic cop pulls you over for rolling through a stop sign. As he approaches your car, you:
   a. Place both hands on the steering wheel and stare straight ahead.
   b. Roll down your window and calmly ask, "What seems to be the problem, officer?"
   c. Place your driver's license on the dashboard next to a twenty-shekel bill.

6. You're at the park with your child when suddenly you spot a crying toddler hurt on the ground. You:
   a. Do nothing. Surely the child's parent or babysitter will come to the rescue.
   b. Look for an approaching adult and, if necessary, yell, "Is this somebody's child?"
   c. Organize a search party to locate some Bamba (the peanut-based puff snack that's every Israeli child's favorite comfort food).

7. Your financial goal for the upcoming year is:
   a. To contribute more into your retirement fund.
   b. That by December 31, you'll have grown your savings by 2 percent.

   c. To invest 10 percent of your salary into your brother-in-law's army buddy's officer's biotech start-up, which is guaranteed to make you $20 million by October.

8. Your El Al flight has just landed at Ben Gurion Airport. You:
   a. Stay seated until the pilot announces that it's safe to stand.
   b. Stay seated until someone else stands, at which point you bolt out of your seat.
   c. Jump up, grab your three roller bags from the overhead bin, and rush to the front of the cabin while the plane is taxiing so you can be first off the plane, first to passport control, first to baggage claim, and first home.

9. At the morning staff meeting, your colleague Daniel presents a slideshow about market trends. You feel the presentation is under-researched and don't agree with most of Daniel's conclusions. You:
   a. Stop by your colleague Ruti's cubicle after the meeting and ask what she thinks of Daniel's presentation.
   b. Stop by Daniel's cubicle and offer feedback according to the "sandwich model": first praise, then constructive criticism, then more praise.
   c. Tell Daniel after the meeting that not since the Yom Kippur War has someone been so ill prepared, and unless he cleans up his act, he, like Golda Meir, should plan to resign.

10. Your sister introduces you to her friend Ron, who has just made aliyah from Brooklyn. You say:
   a. *"Mazal tov!"* (Congratulations!).
   b. *"Baruch haba!"* (Welcome!).
   c. *"Sababa, achi, eizeh kef lecha!"* (Awesome, my brother, how wonderful for you!) ... *"Next time you go to US, you bring me iPhone"* (Please smuggle in an iPhone on my behalf so I don't have to pay full price).

All done?

Great!

Now let's find out how Israeli you are! Give yourself one point for every answer you marked *a*, two for *b*, three for *c*, and total your points to determine your Sabra Score (we'll explain), as follows:

10–20 points. *Yo, eizeh basa* (What a bummer). You are definitely not Israeli! Our best guess is that you live someplace where people are extremely well mannered and overly polite. Probably Canada.

21–25 points. *Kachah kachah, lo nora* (So-so, not terrible). You're still not Israeli, but you're not quite Canadian, either. You're probably from someplace where people are both genuinely nice and emotionally repressed. We're guessing London.

26–30 points. *Teruf* (Madness)! You scored in the top bracket, a clear indication that you are categorically, unquestionably, and quite emphatically...*still not Israeli*!

Wait, what?

That's right. We admit it, the quiz was rigged. None of the answers listed were actually correct. But that doesn't mean the quiz wasn't real – any *true* Israeli would have known the answers were bogus! To understand why, keep reading...

Quiz Answers: What's Your Sabra Score?

Yalla (Let's go)! It's time to see how you should have answered...

1. Stoplight. The light turning green is the first clue that something is awry in this scenario. As any true Israeli knows, the stoplights in Israel don't change from red to green, but from red to both red *and* yellow before finally green. A subtle difference, but critical to what happens next. Secondly, the only Israelis who sit patiently while the car ahead of them dawdles are *freiers* (Hebrew for "suckers"; see chapter 1). A true Israeli would honk *before* the light turns green, a split second after it switches from red to red/yellow, to ensure forward motion by the time the green appears. Israelis live for today – who can afford to sit and waste two valuable seconds of their lives? This requires impeccable timing, which Israelis have down to a science. (They apparently learn it in driver's ed.) It's also why pedestrians should never attempt a last-second dash across the street, because the cars will already be accelerating toward you.

2. **Bar mitzvah.** If you guessed *a* or *b*, we understand. Bar and bat mitzvahs are a really big deal in the US, like miniature weddings, just without the guest of honor's best friend mumbling that he's making a big mistake. For the average non-Orthodox Israeli, however, the bar mitzvah is nothing more than a quick blessing next to the Torah followed by a catered lunch. No $100/hour tutors, no engraved kiddush cup and awkward handshake from shul president Arnie Goldfarb, and definitely no ceremony where your bunkmates from Camp Ethel Rosenzweig stand to light candle number eight. But even if bar mitzvahs *were* as big a deal in Israel, dads still wouldn't dress up. Why? Because most Israeli men don't own suits! When it comes to clothing, it's always Casual Friday – even at your son's bar mitzvah. So the next time you attend one, don't insult the host family by dressing nicer than they do. Instead, dress like the boy's father in jeans, sneakers, and a nice shirt (collar optional).

3. **Wedding gift.** Israelis don't give china, his and hers beer steins, or any of that home furnishing stuff. Engaged couples don't even register, because "that's just *so* American" (i.e., superficial and extravagant). Besides, why give housewares when what the newlyweds *really* need is money to buy a house? In Israel, there's only one wedding gift: cash. Preferably in large denominations (two-hundred-shekel bills), although check, bank transfer, or Bit or PayBox (the Israeli versions of Venmo) will also suffice; if you're really feeling generous, throw in a lottery ticket, too. So why isn't *c* the correct answer? Because in order to give a gift, you need to first estimate the cost of the wedding meal and try to match it! For example, if your cousin's getting married

at the swanky Dan Panorama Herzliya, you're probably looking at a four-hundred-shekel meal, which translates into an NIS 800 gift from you and your plus-one. If, however, she's getting married at a standard, cookie-cutter hall in boring Petach Tikvah, that's more like 275 shekels per plate. It's not unheard of for guests to do the math only after the meal to better assess the quality and value of their salmon entrée.

4. Job search. The key to getting a good job is "Vitamin P," and don't bother looking for it at Super-Pharm. Vitamin P is *protektziah* – slang for "connections," or preferential treatment based on who you know. As explained in chapter 1 (in "Trait #4: *Shivyon*"), everyone is separated by so few degrees that the first thing Israelis do when they need something is figure out whom they know in high places. Good for those already privileged with money and experience, less so for those who could really use a break, but either way, *protektziah* is the grease that lubricates the engine of Israeli life.

5. Stop sign. Over the course of our more than twenty-five combined years living in Israel, we can count on two hands (one each) the number of times we've seen a driver pulled over for speeding, illegal U-turns, or other infractions, and not once, ever, for running a stop sign. In those few instances, we've sometimes even witnessed the drivers *getting out of their cars to argue with the cops* – sometimes yelling, foot stomping, and other stuff that could get you tased in (insert American city) but in Israel are par for the course.

6. The fallen child. Between the two of us, we've got two stories that perfectly illustrate this point. A few years ago, Joel was in the checkout line at a grocery store in the US when he noticed that the toddler in the cart ahead of him had wriggled free from the child restraint and was attempting to stand. The mother, busy searching for her wallet, was oblivious to his imminent fall. "Whoa, careful, sweetie!" Joel said, lifting the toddler up and preventing him from serious injury. Instead of thanking Joel, however, the mother threw a fit at this crazy stranger for daring to touch her darling baby! Joel tried to explain, "The baby stood up. He almost fell. I just – " But the mother thanked him by reporting Joel to the store manager, accusing him of predatory behavior and threatening to sue. Moral of the story: in the US, if you see a child in trouble, *do not, under any circumstances, pick the child up!* Better to let the kid suffer and avoid the hassle of litigation.

Contrast this with Benji's experience on an Israel-bound flight. When the Israeli mother in the adjacent seat had to use the restroom, she handed Benji her infant for emergency babysitting – which he was happy to do! (Even if it was unpaid.) In Israel, kids are in many ways everyone's responsibility. Therefore, the correct answer is to pick up and comfort the crying child until the parent arrives. You won't be sued. (Worst-case scenario, you'll have no trouble finding a Jewish lawyer.)

7. Financial goal. You're kidding, right? Financial *goal*? Not here! Fiscal planning makes sense in countries with normally functioning economies. But in Israel, where salaries are low, taxes high, and gas can cost almost nine dollars a gallon, most families just want to *soger et hachodesh* (finish the month) with minimal debt. Or, more realistically, less debt than the month before. Most Israelis live in overdraft, or *meenus* (from the English word *minus*); the law allows it, and banks encourage it. When it comes to money, Israelis think in terms of the current month, not next month or next quarter, and certainly not next year.

8. El Al flight. *B'emet* (Really)?! El Al?! Their loyalty program is confusing, and they don't partner with other airlines, so it's hard to accrue enough points to go anywhere. Yes, it's ours, and it's safer – they're the only airline in the world for whom part of their USP (unique selling proposition) is the presence of at least one armed plainclothes commando on every flight. And they were the only – we repeat, *only* – airline in the world that continued to fly into and out of Israel in the aftermath of the October 7th attack by Hamas. But what good is a safe arrival if the points you just accrued are only valid on Tu b'Shevat to Bangkok via Alpha Centauri? But airline aside, you, as an Israeli, will be out of your seat the instant the wheels hit the tarmac, if not before. Because really, what are they going to do – turn the plane around and return to Newark because you stood up before the flight attendant gave you permission to stand? Seriously.

9. Staff meeting. When it comes to giving feedback, Israelis don't subscribe to the sandwich model. Nor do they frame criticism as "about the behavior, not the person." Israelis take a direct approach. If Daniel gives a bad presentation, there's only one person to blame, and that's Daniel! You might as well tell him so immediately, in the staff meeting. It's not hard to do: just cut him off mid-sentence, say exactly what you think, and don't dare let him respond while you're interrupting. If he's a true Israeli, he can take it.

10. The friend who moved to Israel. When Israelis meet *olim chadashim* (new immigrants), their first words are often a disbelieving "Why?" As in, why on earth would you leave the United States, a land flowing with organic milk and honey, where you can buy a home with no money down, there's a Target on every corner, and you can return a single dirty sock to a department store without a receipt, no questions asked? The not-so-subtle implication is that you, the immigrant, are a complete and utter doofus for leaving a life of luxury and convenience when you could have sold Dead Sea mud at the Mall of America or worked for Moishe's Moving Company in Queens.

If you didn't realize it before, now you know: Israelis ain't like the rest of us.

But why? What is it that makes them so darn, you know... *Israeli*? And how did they get that way?

Grab yourself some sunflower seeds, pop open a Goldstar beer (or Nesher Malt if you're under twenty-one), and get ready for an exhilarating journey into the mind of the modern Israeli. We're about to explore why Israelis think, act, and behave (and misbehave) as they do...

CHAPTER 1

The Israeli Psyche

IMAGINE A PLACE WHERE PEOPLE SAY WHAT THEY MEAN AND mean what they say. No subtext, no political correctness. If your feelings get hurt, that's your problem. Stop being so sensitive!

Now imagine that in this place, there are no rules. Or actually, there are rules, but nobody thinks the rules apply to *them*. "Rules are for idiots who can't think for themselves, but *me*? I've got this!"

Finally, imagine feeling that everybody who lives in this place knows everybody else – like you're all one big family. The bank teller, the coffee barista, and the friendly old cabbie who drives you to the airport may be strangers, but you can strike up a conversation as if you've known each other for years.

Well, guess what. This place you've been imagining? It exists. Welcome to Israel.

To truly make sense of a place, you need to understand the people who live there. We therefore begin our exploration of all things Israel with an in-depth look at the Israeli personality.

Throughout this opening chapter, we turn to two people for their insights, each of them experts in their own way: Dr. Tamar Katriel, anthropologist at Haifa University and one of the country's authorities on what she calls "the Israeli ethos"; and Moshe Samuels, Jewish educator and creator of the Israeli TV show *The New Jew*, in which well-known comedian Guri Alfi toured Jewish communities across the

United States to explore the differences between American and Israeli Judaism. With their help, we'll discover not just how Israelis think, but *why* they think and act as they do.

Meet the Sabra

When you ask Americans and other Anglos what they think about Israelis, the first thing they'll often say is "rude."

When we first made aliyah, we might have agreed. But the longer we live in Israel, the more we've come to realize there's actually no universally agreed-upon line in the sand demarcating rudeness; there's only the question "Was someone offended?" And if so, that says as much about the person whose feelings were hurt as it does the agent of the (supposedly) offensive behavior.

The fact is, Israelis generally don't get insulted by many of the behaviors that bother Americans and other Westerners. You can take a buddy's french fry without asking, tell a friend she looks tired, or ask someone you just met how much she pays for rent (if she hasn't already volunteered that info herself).

So if you bought this book hoping to learn, once and for all, "Why are Israelis so outspoken and rude?!" we'll do our best to explain – but keep in mind, that's only part of the question. The other – and perhaps more interesting – part is *why* these behaviors that we Westerners perceive as rude are not only acceptable in Israel but actually *laudable*.

Joel learned this the hard way a few years ago while buying school supplies for his kids. Like many towns, Ra'anana only has one bookstore that sells the requisite school textbooks. Teachers release the book list two or three days before classes start, and from that moment, it's a mad dash to gather the necessary supplies. Imagine Black Friday, without the Christmas decorations.

Foreseeing the chaos and not wanting to spend hours in line, Joel did what any conscientious Midwestern American might do: he showed up early – seven a.m., to be precise, an hour before opening – to make sure he'd be the first in line.

Indeed, he was. By eight o'clock, a long line had formed behind him, snaking around the corner. Victorious, Joel gave himself a well-deserved pat on the back.

A minute later, the store manager unlocked the door, and everyone who'd been waiting patiently behind Joel steamrolled past him inside.

"*Mah koreh poh?*" (What's happening here?), Joel exclaimed. "I was first in line!"

The manager shrugged. "That's how it works in this country," she said, clearly pegging his attitude as a foreigner's. "If you don't like it, don't live here."

That's the sabra mentality.

An Israeli who's born in Israel is called in Hebrew a *tzabar* (cactus). In this book, we'll use the English word, *sabra*.

Sabra is more than just a nickname, it's a metaphor. The idea is that Israelis are like cacti – thick-skinned and prickly on the outside, but soft and sweet on the inside.

As it turns out, the sabra label is older than the country itself. The term originated in the early 1900s to describe the children born to the first wave of immigrants, the *chalutzim* (pioneers), who came to then Ottoman-controlled Palestine during the First Aliyah. These first-generation sabras took pride in being outspoken, tough, and selfless in their commitment to an ideal larger than themselves – the building of the State of Israel. A key component of their new identity was what they were *not*: timid, frail, and helpless (which their nebbish Jewish ancestors were, from their perspective).

Israelis today still pride themselves on these same cactus-like qualities. Power, strength, and brutal honesty (the outer part of the cactus) are among the most admired traits in Israeli society; so are hospitality, camaraderie, and self-sacrifice on behalf of the common good (the sweet inside).

Spend a month or two in Israel, and you'll see examples of prickly-sweet sabra-ness everywhere. It can happen in the span of a minute: a falafel shop owner who barks, "Just a second, can't you see I'm busy!"

when you try to order, and then, thirty seconds later, hands you a piping-hot falafel ball straight from the oil and says, "For you, *motek* [sweetheart]. Now, what would you like?"

That said, "prickly sweet" is less about what you'll encounter in any single interaction and more about the extreme range of behaviors you're likely to experience over time. Someone could light up next to a "no smoking" sign, scam you out of ten thousand shekels, and drive the wrong way down a one-way street in a school zone ... but he'll also pull over to help you change your flat tire and invite you to Shabbat dinner without knowing your name. On both the individual and group levels, the range of behaviors that Israelis exhibit is wide. Or as one sociologist we spoke to put it, "The bell curve of Israeli behaviors has very long tails."

What do we mean by "long tails"? That the average Israeli is capable of more extreme behaviors in both the positive and negative directions – say, being both more rude (or, what *we* might consider rude) and more generous (again, from our perspective) than you're likely to see on a day-to-day basis relative to a typical Westerner.

For example, about a year ago, Joel called the customer service department of his new Wi-Fi provider. Here's an actual snippet from that conversation:

> JOEL: I'm having trouble connecting the Wi-Fi to my smart TV.
> REP: Why would you buy a smart TV if you're not smart enough to use it?

Ouch!

But Joel knew that if he could just keep the guy on the line long enough, that same customer service rep would eventually help him.

And he did. The rep consulted with technicians, spoke with his manager, and contacted managers of other departments. When none of that worked, the rep drove thirty minutes out of his way to Joel's apartment to try to fix it himself. He did – and refused a tip.

But the story doesn't end there, because on his way out the door, the rep, who by now had introduced himself as Lior, got to talking with

Joel's wife, Dorit, about the *shakshuka* cooking on the stove. They spoke about favorite recipes and the merits of Moroccan versus Yemenite cooking, which then led to the discovery that Lior, like Dorit, was half-Yemenite. Just like that, Lior invited Joel's family to his mother's house for Shabbat *jachnun* – a traditional Yemenite Sabbath dish that, Lior said proudly, his mom makes from scratch.

An extreme behavior in the other direction.

Still, the cactus metaphor can only explain so much. Cutting in front of someone in line at the bookstore, that's prickly. Driving the wrong way past a busy school, that's downright reckless! How do you explain *that*?

Nor does "cactus" tell us why an Israeli will take the shirt off his back to help you, literally. A few years ago, on a hot July day, Benji was standing at a bus stop when the elderly woman next to him fainted and gashed her forehead on the pavement. While Benji looked for water, others cradled the woman in their arms and called for paramedics, and one thirty-something young man took off his shirt and pressed it to the woman's head to stop the bleeding.

Gross? Kind of. But a genuine mitzvah and quintessentially Israeli.

There are countless reasons that any person is the way he or she is. But based on our own experiences living here in Israel, as well as interviews with professors, sociologists, and other people much smarter than us, we've concluded that there are five cultural values you absolutely must be familiar with if you want to understand Israeli society.

In the following chapters, we'll introduce these traits one by one and explain the origins of each. We begin with one of the most important, one that every Israeli knows (including children) but which, somehow, few non-Israelis have heard of…

Trait #1: *Freier*

If you went to Jewish summer camp, you probably learned a bunch of supposedly important Hebrew words such as *shalom* (peace/hello/

goodbye), *ochel* (food), and *sheket b'vakashah* (please be quiet – a useless phrase, since no Israeli would *politely* ask you to shut up).

The one word you never learned but should have is *freier*.

Originally Yiddish, *freier* (pronounced "fryer," as in one who fries) is Hebrew slang for "sucker" or "dupe," only worse – because inherent in being a *freier* is the idea that you *allowed* yourself to get taken advantage of.

Understand this: in Israeli society, the absolute worst possible thing a person can be is a *freier*!

We're not kidding. The mere thought that someone might be trying to screw them over is enough to drive Israelis nuts. Which is why they will do anything to avoid it.

For example: You're on the highway, stuck in traffic. You glance to your right and notice that the car next to you has put on his blinker and – *are you kidding me?* – is trying to cut in front of you, into *your* lane. So what do you do? Simple: you inch forward and block him – because only a *freier* would allow some jerk to cut him off. And you are not a *freier*!

Example two: You're on the highway, stuck in traffic. You glance to your left and notice that traffic in this lane seems to be moving a bit faster. So you put on your blinker and try to merge, when – *are you kidding me?* – the idiot next to you inches forward to block you from cutting in. So what do you do? Simple: you accelerate and try to merge before he can cut you off – because only a *freier* would allow some jerk to stop him from going where he wants to go. And you are not a *freier*!

Israel is notorious for its aggressive drivers and high number of traffic accidents. Aversion to being dubbed a *freier* is one of the main reasons.

For Israelis, the obsession with not being taken advantage of isn't restricted to the roads; it permeates every element of society. TV and radio commercials declare that "only a *freier* would shop somewhere else." Politicians rally crowds by asking, "Are you with me, or are you a *freier*?" Students argue with their classmates over who will be the *freier* of the class project, lest they be the one who stays up all night coloring, cutting, and pasting while the others sit back.

The concept is such a big deal that actual scholars have dedicated their careers to studying it – most notably, Linda-Renée Bloch of Tel Aviv University. In a 2005 paper, she wrote:

> The negative perception of being a *freier* is so strong, that its antithesis – the imperative not to be a *freier* – has become a value in and of itself, engendering acts that demonstrate how one is not a *freier*, and these frequently overshadow what is, at least theoretically, the principal goal of the interaction.[1]

When Israelis find themselves in conflict, what's more important than coming out on top is *not* coming across as a *freier*. The next time you see two Israelis fighting over, say, a parking spot, keep in mind that what they're actually fighting about isn't parking, but their *identity*. What's at stake is nothing less than their sense of self – neither wants to be perceived as the spineless wimp who caves in. So they double down and fight to the death, sometimes literally: Israelis have been known to kill each other over parking spots. (The Middle East is a dangerous place.)

Their preoccupation with not being a *freier* leads to other uniquely Israeli behaviors:

- **Ignoring rules.** Only a pushover would allow someone else to tell him what to do. This is why you'll see motorcycles zooming down the sidewalk at full speed and people smoking next to "no smoking" signs. As one immigrant friend of ours so aptly put it, "In Israel, rules are merely suggestions."
- **Disrespecting authority.** To show deference to authority is to acknowledge that someone else knows better than you. Only a *freier* would think that!

---

1. Linda-Renée Bloch, "Who's Afraid of Being a *Freier*? The Analysis of Communication through a Key Cultural Frame," *Communication Theory* 13, no. 2 (May 2003): 125–29.

- **Arguing.** Only *freiers* take no for an answer. That's why Israelis will continue to argue even after it's clear they've lost the argument.
- **Refusing to apologize.** In the Israeli mindset, "I'm sorry" equals "You have the upper hand." Part of not being a *freier* is never showing weakness, ever.
- **Cutting corners.** Only *freiers* do stuff "by the book." True Israelis innovate, improvise, and find faster, cheaper ways of getting the job done. This mindset has led to incredible high-tech innovation and military success – but also, unfortunately, to horrific accidents. (Search "Meron tragedy Israel," "synagogue bleacher tragedy Israel," and "construction site tragedy Israel" for stories about how cutting corners led to disaster.)

The concept of *freier* isn't just about self-perception; it has a powerful social component, too. Friends and family will give you a hard time if they think you've been a *freier*. The *freier* concept therefore colors just about every interaction in which something is at stake – between customer and merchant, worker and boss, husband and wife, and any other relationship where there's something – or the perception of something – to be lost or gained.

Believe it or not, "*freier*" also helps explain why it's so hard to achieve peace in the Middle East. One of the strongest values in Arab culture is the preservation of honor, commonly known as "saving face." Between that and Israelis not wanting to look as if they've been taken advantage of, you can see why it would be so difficult to reach a compromise. No matter how good the terms are for each side, there's always the chance you'll be perceived as the chump who gave up too much and let the other side get the better of you.

Rumor has it that at the Camp David peace talks in 2000, Israeli prime minister Ehud Barak and Palestinian Authority president Yasser Arafat both refused to walk through the door into the negotiating room until the other entered first; each motioned for the other to go in until, finally, President Clinton pushed both leaders through the door at the same time. It sounds childish, but in the context of deeply ingrained cultural values, it makes sense: entering a room first is a signal of inferiority.

This is what the rest of the world tends to miss – that the past and present Arab-Israeli conflicts aren't just about security, settlements, and 1967 borders. They're about perception: who looks strong, who looks frail, who saves or loses honor.

And who comes out looking like a *freier*.

Of the five values we explore in this chapter, "*freier*" may be the most difficult to unpack. Below, we present overviews of four theories.

Sheep to the Slaughter

"Victimhood is one of the strongest elements of the Israeli ethos," says Dr. Tamar Katriel, the Haifa University anthropologist who's dedicated much of her career to studying how Israelis behave and think.

You don't have to be a historian to know what Katriel is talking about. Exiles, expulsions, inquisitions, pogroms... For as long as there have been Jews, there have been people who persecuted Jews – from the biblical Amalek to Nazis to neo-Nazis. It culminated in the Holocaust, when six million Jewish men, women, and children were murdered while the world watched in silence. As well known as those horrors are, the scale of the trauma is still mind boggling – all the more so because the hatred of Jews continues still today.

The Israeli psyche is deeply scarred by the collective memory of Jewish suffering – and the Holocaust in particular – for a few reasons. First, almost all Israelis are descendants of someone who survived something terrible. In 1948, of the eight hundred thousand Jews in Israel, a hundred thousand were Holocaust survivors who brought with them their trauma, mistrust of authority, and stark us-against-them worldview – a mindset they then passed on to their children, grandchildren, and so on. Not long after European Jews were being rounded up and forced into gas chambers in Europe, seven hundred thousand Middle Eastern and North African Jews fled or were expelled from Iraq, Egypt, Turkey, Morocco, Yemen, Libya, Iran, and other Arab countries. When these refugees got to Israel, they had to live in tents or other temporary housing and were discriminated against by their Ashkenazi neighbors. Years later, waves of Russians and Ethiopians made aliyah, fleeing persecution, only to face discrimination upon arriving in Israel. Many of today's *olim* do so to escape anti-Semitism in France, Turkey, and other countries.

The national mentality in a few words? "Someone's trying to get me." Not exactly "land of the free and home of the brave."

In the early years of the state, many Israelis regarded Holocaust victims as the ultimate *freiers* – standing naked in the cold, hands in the air, marching like sheep to the slaughter. It's a harsh view, but, as Jewish educator Moshe Samuels explains, a necessary one. "The *chalutzim* who founded the state *had* to think that way in order to believe that the new country could survive," he says. "Now, if they had been the ones rounded up in Europe, would they really have acted differently? We don't know. But generations grew up learning to negate the Diaspora, as if to say, 'Look what happens when you don't have a Jewish state.'"

Deep down in their consciousness, Israelis have made a vow: now that we have a place of our own, we won't be bullied, beaten, or marched like sheep to the slaughter.

We will not be *freiers*.

"The War"

The *freier* concept isn't just a response to two thousand years of persecution. Israelis are keen to never be duped, suckered, or swindled for a more contemporary reason: "the war."

Which war, you're wondering. Depends on the Israeli's age.

For the senior citizen sitting next to you on the bus, cane in one hand and shopping bag filled with groceries on the floor, "the war" might be the Yom Kippur War in 1973, when the country was attacked by Syria and Egypt on the holiest day on the Jewish calendar, and more than twenty-five hundred Israelis were killed (in percentages, the equivalent of 250,000 Americans dying today), among them her father.

For the thirty-something next to you in the café, typing away on his laptop, "the war" might be Operation Cast Lead (2008) or Protective Edge (2014) – or both – when he fought as an infantry soldier in Gaza and lost two friends.

For the young mother in the park, pushing her daughter on the swing, "the war" is Ason Hamasokim (the Helicopter Disaster) of February 1997, when two IDF transport helicopters collided on the way to Lebanon, and seventy-three Israeli soldiers were killed, her older brother among them.

And for her six-year-old daughter in the swing, pigtails flying behind her head, "the war" is being woken up at two a.m. by the *azakah* (siren), running to the bomb shelter in her pajamas, and feeling the building shake from the thunderclap of Iron Dome overhead, during the October 7th war that began in 2023 and continues as of this writing – a memory that returns in her nightmares.

Beyond just the sheer number of wars, military operations, kidnappings of soldiers and civilians (including children), and terror attacks that the country has endured, Israelis' relationship with war is uniquely personal in three ways:

- **When.** It's been said that wars in the Middle East are like buses: if you miss one, there's sure to be another one coming soon. Indeed, Israel has gone to war, on average, once every six to eight years (and about once every three years since the year 2000), meaning not a single Israeli has ever grown up knowing total peace. Israelis talk about war like Americans might discuss the weather: "I heard things are heating up in the north."
- **Where.** With the exception of Pearl Harbor and the September 11 attacks, America's modern conflicts always happen in far-off places that its average citizen will most likely never visit. For Israelis, meanwhile, war is local: all conflicts happen on Israeli soil or just over the border. The southern city of Sderot now has a missile-proof indoor playground so that kids have a place to play during conflicts.
- **Who.** Israel's wars aren't fought by anonymous soldiers whose names and faces the rest of the population will never know; they're

fought by people's sons and daughters, brothers and sisters, neighbors and friends. Every Israeli knows at least one person who was killed in a war, military operation, or training. Next time you're in Israel, notice how many playgrounds are named after nineteen- and twenty-year-olds whose lives were lost too soon.

The persistent and personal nature of war, along with the threat of imminent attack, both perceived and real, colors how Israelis see and react to the world. Their antennae are up, and the moment they sense even a *hint* of someone trying to get the better of them, they spring into survival mode: *"I won't be made a freier!"*

The Shuk

One source of the *freier* concept also happens to be the simplest: it's the Middle East.

"Israel is a Middle Eastern country with a *shuk* mentality," Samuels says. "In the *shuk*, buyer and seller are constantly competing and bargaining for a better deal. The starting assumption is, 'You're trying to screw me.'"

While this kind of haggling may have originated over carpets and cucumbers in the *shuk*, it didn't remain there. Israelis negotiate over everything from the cost of internet service to phone plans and washing machines (something you don't often see at Best Buy). It's also transcended the world of commerce. "This *shuk* mentality has trickled down into every type of interaction where someone has something to gain or lose," Samuels says.

So while in general, people tend to trust strangers until they have a reason not to (an idea Malcolm Gladwell explores in his 2019 book *Talking to Strangers: What We Should Know about the People We Don't Know*), in Israel the opposite rings true: "This guy is gonna try to dupe me, so I need to watch my back and not be a *freier*."

The Kibbutz

Is it possible that, in addition to persecution, war, and fear of getting suckered into a bad deal, *"freier"* might actually be related to a more positive social trait?

Katriel believes so. For her, the *freier* concept is a reaction to one of the core principles on which the country was founded: collectivism.

"In Israel, our sense of collectivism and contributing to the common good is so strong that at some point people start asking, 'Hey, what's in it for me?'" she says.

One of the earliest examples of collectivism in Israeli life is the kibbutz – communal settlements where residents ate together, shared property, and all income went into a communal pot. The word itself, *kibbutz*, means "gathering together"; on most kibbutzim, private property and income were not allowed. Although the country has fewer kibbutzim today, and those that do exist have to varying degrees been privatized, the old-school together-for-the-common-good kibbutz ethos is still a part of the Israeli social contract.

A more contemporary version of collectivism is mandatory military conscription. At eighteen, women are drafted for two years of service in the IDF, men for two years and eight months – the longest mandatory enlistment period of any democracy in the world. After that, many men (and some women) continue to serve in *miluim* (reserve duty) into their forties, requiring them to leave their families and jobs for days or even weeks at a time each year to train and, if necessary, go into battle.

And then there are taxes. It's not just that taxes in Israel are high; it's the perception among many citizens that not all are paying their fair share. There's something to it: according to Professor Daniel Ben David of the Shoresh Institution, 92 percent of income tax revenue comes from 20 percent of the population; a whopping 50 percent of Israelis don't earn enough to pay income tax at all. Many of them, Ben David says, fail (on purpose) to declare large amounts of cash income they earn *mi'tachat la'shulchan* (under the table). Another issue is the Haredi community, a population of 1.2 million who generally do not serve in the army and under-participate in the labor force yet enjoy government stipends for each of their (on average) 6.5 children, plus money for yeshiva studies. Although more Haredim are entering the labor market, the *freier* mentality is, as always, mostly about perception – and the perception is that not everyone is anteing up.

According to Katriel, *"Ani lo freier!"* (I am not a *freier!*) is how Israelis keep the tension between collectivism and individualism in check. Israelis are happy to pitch in, but playing by the rules when others don't would be foolish.

Like the great Rabbi Hillel said, "If I am not for myself, who will be for me? And if I am only for myself, who am I?" (Ethics of the Fathers 1:14).

Or, as Katriel puts it: *"'Freier'* is our way of saying, 'I'm happy to give, but only to a point.'"

So if you ever get caught trying to "pull a fast one" on an Israeli, they'll call you on it. *"Mah ani, freier?"* (What am I, a *freier?*), they'll exclaim.

As to *why* they'll call you on it, rather than slink away and avoid the confrontation... for that, we need to explore the second of our Israeli traits, *dugriyut.*

Trait #2: *Dugriyut*

Have you ever asked someone for an opinion or feedback and wondered if the answer is how the person *really* feels?

Good news: you'll never have to worry about that in Israel!

Of all the elements of the Israeli character, the one most non-Israelis are familiar with (even if they don't know it by name) is *dugriyut* (direct speech, pronounced DOO-gree-oot). Israelis tell it like it is, no sugarcoating, no subtext. Just the cold, hard truth, even if it hurts. Heck, *especially* if it hurts. If Israelis are cacti, their unfiltered *dugri* speech is the thorns.

When people say Israelis are rude, there's a good chance it's *dugriyut* that they're reacting to. For starters, it's salient: when the person you're talking to speaks bluntly, unfiltered, you feel it. And unlike other Israeli traits, *dugriyut* is personal – at least, that's how it feels, even if it wasn't meant to be.

Like the word *freier, dugri* is an import from another language – in this case, Arabic. But the word means different things in each culture. In Arabic, *dugri* means "straight"; the phrase *"Dugri, dugri!"* is often used when giving someone driving or walking directions ("Go straight and keep going straight!"). In Arab society, *dugri* speech means telling the truth even if it could incriminate you: *"Dugri,* I stole the cookie from the cookie jar." Contrast this with Israel, where *dugri* is more about sharing your honest opinion (and here's where Westerners cringe): *"Dugri,* I stole the cookie from the cookie jar, and it was the worst cookie I've ever tasted. Who taught you how to bake?"

For sensitive Americans like us, this in-your-face speech can take some getting used to. Being told that you've put on weight or that your hair's going gray – that can sting! Indeed, new immigrants commonly say that Israelis' unpolished, straight-to-the-point communication is one of the hardest parts about life in their new country.

But eventually, you start to appreciate the positive aspects of *dugri* speech, like not having to waste time and energy trying to figure out what the other person *really* means. Why? Because they just told you!

Another upside of straightforward speech is the effect it has on relationships. Because of *dugriyut,* friendships in Israel get very deep, very fast. Israeli men, regardless of age, will address each other as *"achi"* (my brother). People open up from the start, and no topic is off the table, so you get to know each other in full more easily and authentically than you would elsewhere. Benji experienced this himself as a new immigrant. Before he owned a car, he often relied on carpool and rideshare apps to get around the country. Though none of the drivers and passengers knew each other, he was struck by how quickly they would hit it off and interact like old friends. Within minutes, they knew each other's names, where they lived, their job or area of study, political affiliation, where they served in the army, marital status, and which country they supported in the Eurovision song contest. At the end of the ride, they sometimes exchanged numbers and made plans to stay in touch – which they did, because unlike Americans, Israelis only say "We should get together" when they *mean* it!

Several of our Israeli friends shared the flipside of this story from their time living in the States: when a friend or colleague would say, "Let's do lunch!" the Israeli would (a) inevitably realize that this lunch meeting would never happen, or (b) call their bluff by whipping out their calendar, leaving the American speechless.

Meanwhile, Joel's auto mechanic, Avner, knows exactly how much money Joel earns in an average month, the amount he pays for rent, and that he and his wife went to marriage counseling in 2015. And, of course, Joel knows the same about Avner, including the *reason* he and his wife went to therapy – when Avner was four, his dad walked out on the family, and Avner has struggled with intimacy ever since.

How much do you know about *your* auto mechanic?

When Israelis living abroad describe what they miss most about Israel, they often say it's the depth and authenticity of the friendships. A product, in large part, of *dugriyut*.

For anthropologist Dr. Tamar Katriel, *dugriyut* is about speaking truth to power. She cites the story of IDF Colonel Eli Geva, a high-ranking combat officer who was ordered to lead his troops into Beirut during the 1982 Lebanon War but refused. "Geva was called to appear before Prime Minister Begin," Katriel says. "Asked to explain himself, Geva said, 'When I looked into my binoculars, I saw children playing.' He was relieved of his post, but he'd spoken his truth."

Not that this is a recent phenomenon. In her view, speaking truth to power has deep roots in Jewish tradition, exemplified by the late prophets of the Bible. "When Jeremiah says to God, 'You deceived me!' that's *dugri*. Other prophets do it too. They question God, the ultimate authority figure. And tell it like it is."

Where does this straight-to-the-point speech come from, and how did it come to define how Israelis communicate?

To answer that, we're going to introduce one of the most important ideas motivating Israel's founding: Naye Yidn, Yiddish for "New Jews."

When Theodor Herzl, A.D. Gordon, and other visionaries dreamed of a Jewish homeland, they envisioned not only that Jews would have a country of their own, but one populated by an entirely new paradigm of Jew. In their minds, the Naye Yidn would be strong, brave, and confident, a complete turn from the cowering, overly apologetic Alte Yidn (Old Jews) of the past. The Naye Yidn would be different in more ways than just character: Instead of Yiddish, they would speak Hebrew. Instead of working as money lenders, they would work the fields with their bare hands, men and women tilling the soil side by side for the collective good. They wouldn't let anyone push them around, either. Babylonians, Cossacks, Nazis? *Tishkechu mi'hem* (Forget about them)! Naye Yidn would kick ass and take names.

Best of all, New Jews could say and do whatever they wanted. Old Jews had to watch what they said, lest they upset the authorities. But in their new state, Jews would be the authority! As Katriel explains in her book *Talking Straight*:

> The Israeli Jew was to be everything the Diaspora Jew was not. In communicative terms, this implied the rejection of ways of speaking associated with European genteel culture and Jewish Diaspora life in particular.[2]

For thousands of years, Jews had to talk nice in order to survive. But with the establishment of the State of Israel, they finally held their fate in their own hands. Self-determination gave them the freedom to speak their minds without fear of getting lined up, beaten, and killed. For the first time, Jews had the freedom to fully express themselves in all areas of society, even to criticize people at all levels of power. If there's one place where the Jews actually do control the media, it's Israel.

In the earliest days of the country, *dugriyut* was more than just a revolutionary way for Jews to speak – it was also a necessity. Those pioneers had swamps to drain and fields to hoe, and the only way to do it was with direct, efficient communication. You can't beat around the bush when you've got to plant an acre of them by sundown.

When Jewish teenagers and other tourists discover Israel and find

---

2. Tamar Katriel, *Talking Straight: Dugri Speech in Israeli Sabra Culture* (Cambridge: Cambridge University Press, 1986), 17.

themselves smitten by the people, it's often the New Jew aspect of Israelis they're enamored with. For the first time in their lives, they meet what we might call Jews 2.0: sabras who are outspoken, confident, and behave differently from most Jews they know back home. They also look different – they're of every skin color imaginable, from countries you may or may not be able to find on the map. And as you might expect from such a multicultural people, they even *eat* differently, enjoying spice-filled delicacies like *kubbeh* and *mujadara* that are nothing like the bland gefilte fish and roasted chicken their Ashkenazi families ate growing up. (Ashkenazim comprise a minority of Israeli Jews; you'll probably be shocked to hear Israel has very few bagel shops.) By the way, don't even think about eating gefilte fish at a Moroccan Passover Seder – instead, enjoy Savta (Grandma) Shula's famous *chraime* fish recipe.

With a different gene pool and an experience that forked off the road of Jewish history around 1948, Israelis are in so many ways a different people from their Diaspora Jewish brethren – a people who present a breath of fresh air for Diaspora Jews who are tired of the Alte Yidn stereotype.

The New Jew creator Moshe Samuels says *dugriyut* is still the foundation of how Israelis communicate. And in case you're wondering, the answer is no, Israelis are *not* offended by communication just because it's blunt or unfiltered. "Sure, you can choose to avoid political arguments and tough questions in conversation," Samuels says. "But then there's an elephant in the room that creates distance between you. Better to be direct – when you are, it brings you closer. It means you can speak openly with that person without being afraid that it will hurt your relationship."

Trait #3: Chutzpah

An IDF officer puts on a dress and heels, sneaks into Lebanon, and assassinates a Palestine Liberation Organization (PLO) leader to avenge the Israeli athletes murdered at the 1972 Munich Olympics – and is then evacuated back to Israel, unscathed.

Eizeh chutzpah! (What chutzpah!)

An Israeli teenager dreams of flying F-15s in the Air Force but can't because she's a woman – so she sues the IDF, takes her case all the way to the Supreme Court, and wins.

Eizeh chutzpah!

The owner of a rental car agency hands you the keys to your car and, as you buckle up, says the tires are low and you're low on gas, so you'd better go to the gas station across town and take care of it before hitting the road.

Eizeh chutzpah!

The characters in these stories are all real. Prime Minister Ehud Barak was once a covert officer in drag. Alice Miller, a sixteen-year-old *olah* (immigrant) from South Africa, broke the glass ceiling for dozens of women who would eventually earn Air Force pilot wings. Sammy, owner-operator of a local rental car agency (name changed so we don't get sued) made Joel fill his car with gas and inflate the tires, delaying his day trip to the Golan Heights by forty minutes, losing Joel's business for good, and making himself a major *chutzpan* (a male who exhibits chutzpah; *chutzpanit* for a female).

These anecdotes exemplify the best and worst of chutzpah, the engine that powers the vehicle of Israeli behavior.

If you're familiar with only one type of chutzpah, it's likely the type thought of as gall, nerve, or unbelievable audacity that tends to evoke an element of cringe. Licking your plate clean and then demanding a refund from your waiter because your steak was "just so-so" – that's chutzpah. Obnoxious, self-centered, and crossing the line, leading your mortified friends to face-palm and say, "Oh, jeez, I cannot believe he just did that!"

Israel, however, also has plenty of the chutzpah exhibited in the first two stories above. Licking your plate clean, marching into the kitchen, and telling the chef that he simply *must* hire you as his protégé and that you won't take no for an answer – now, that's the chutzpah Israelis appreciate. The kind inspired by phrases such as "no guts, no glory," "what's the worst that could happen?" or "you only live once!"

Model this kind of chutzpah and enjoy as your friends look on in total amazement.

Not only a tool for impressing bystanders, chutzpah also describes the courage to find solutions when the clock is ticking, against the odds, even if – no, *especially* if – you make it up on the spot. Israelis are all about improvisation and taking chances. According to Dan Senor and Saul Singer, authors of *Start-Up Nation*, chutzpah is one reason for Israel's success in the tech industry. Young entrepreneurs have the chutzpah to believe that their wacky start-up ideas are plausible; entry-level employees have the chutzpah to speak up at staff meetings and tell management how things could be done better. And best of all, their bosses actually listen. (They may not take the advice, but they at least listen.) Employers aren't threatened by criticism; they encourage it, because they know that innovation results from workers possessing the chutzpah to question the status quo and conventional wisdom aloud.

Witness Woody Allen nervously kvetching in any movie to see what happens when a stereotypical neurotic Old Jew thinks too much. Paralysis by analysis. When the clock is ticking and the enemy is at your doorstep, who has time to think? A key to good chutzpah is turning down the voice of your inner critic. Hold on – turn it *down*? Israelis don't even *hear* this voice, which makes it easy to approach that chef, ask out a total stranger on the street, or tell the prime minister what you really think. Israelis who demonstrate bold, courageous action are revered with glowing terms such as *gever* (man), *gibor/giborit* (hero [male/female]), *kli* (tool), and *totach/totachit* (cannon).

If, however, someone refers to you as a *chutzpan*, don't be proud; they think you've exhibited the bad chutzpah – selfish, tasteless behavior that goes too far. Like Sammy the rental car agent. Or the young soldier who requests sick leave to recover from an injury at home, only to post Instagram photos of him and his girlfriend at the beach.

Still, we take notice. Because even as we say to ourselves, "What a *chutzpan*!" we secretly wonder if maybe *we* should be sneaking off too. "If I could but don't," we worry, "does that make me a *freier*?"

What a conundrum.

Anyway, it's easy to see why Israelis value chutzpah as much as they do: the country literally would not exist without it.

Theodor Herzl, father of modern Zionism, famously said, "*Im tirtzu, ein zo agadah*" (If you will it, it is no dream). Ever since, the country's pioneers and modern Israelis alike have had the chutzpah to both dream the impossible and pull it off. In fact, Israel's history is a series of one chutzpah-inspired exploit after another. As you know, of course, not all acts of chutzpah are good. Let's review some classic moments of chutzpah throughout history. Decide for yourself whether the protagonist in each is a *totach/totachit* or a *chutzpan/chutzpanit*.

- **1944.** Eighteen-year-old Hungarian poet Hannah Senesh immigrates to British Mandate Palestine, joins a kibbutz, volunteers for the Haganah (precursor to the IDF) as a paratrooper, and, in a secret mission to aid Allied forces, jumps behind enemy lines into Yugoslavia.

- **1948.** David Ben-Gurion declares Israel's independence and states, "We extend our hand to all neighboring states and their peoples in an offer of peace and good neighborliness" – despite knowing that doing so will lead to direct and imminent attack by Arab nations.

- **1949–1950.** In a dangerous operation, Israel airlifts fifty thousand Yemenite Jews to Israel, freeing them from anti-Semitic persecution at the hands of the Yemenite government and drastically improving Israel's gene pool (i.e., upping our chances of one day producing a supermodel and potential Eurovision winners).

- **1962.** The Mossad sends a spy named Eli Cohen behind enemy lines under the identity of Syrian "Kamel Amin Thaabet," since portrayed by a different Cohen (Sacha Baron) on the Netflix show *The Spy*. Claiming he wants to protect Syrian soldiers from the sun, Eli Cohen plants trees on Syrian military fortifications throughout the Syrian Golan Heights, providing soothing shade for troops and unmissable targets for Israeli Air Force pilots in the event they'd ever need to strike.

- **1973.** When Prime Minister Golda Meir tries to convince Jewish-American Secretary of State Henry Kissinger to put Israel's interests first during the Yom Kippur War, he responds, "Golda, you must remember that first I am an American, second I am secretary of state, and third I am a Jew." Golda's reply? "Henry, you forget that in Israel, we read from right to left!"

- **1976.** Operation Thunderbolt: IDF commandos led by Lieutenant Colonel Yonatan Netanyahu fly more than twenty-five hundred miles to Uganda, sneak a Mercedes into the Entebbe Airport past Ugandan troops, and launch a successful hostage rescue mission despite not knowing the layout of the airport or the exact whereabouts of the passengers.

- **1981.** Operation Opera: Ignoring inevitable worldwide condemnation and against enormous odds, Israeli Air Force jets fly twenty-two hundred miles through enemy air space to Iraq, where they destroy Saddam Hussein's nuclear reactor in Osirak. To help them escape in the event of capture, each Air Force pilot travels with a satchel of Iraqi cash.

- **1988.** Right-wing nationalist politician Rehavam Ze'evi establishes the Moledet Party, advocating the population transfer of Arabs out of the West Bank and the Gaza Strip, yet continues to go by his lifelong nickname "Gandhi."

- **1997.** Four women from northern Israel, all of them mothers of combat soldiers, launch the grassroots Four Mothers protest movement to speak out against the IDF's presence in Southern Lebanon. Despite the lack of powerful female voices in military matters, the Four Mothers' campaign led to the unilateral withdrawal from Lebanon three years later.

- **1998.** Israeli start-up Given Imaging designs a groundbreaking camera-inside-a-pill whose data can be collected only after being retrieved from the subject's feces.

- **2007.** Supermodel Bar Refaeli announces her move to America by saying she has no regrets about avoiding army service because

"Why is it good to die for one's country? Isn't it better to live in New York?"[3]

- **2018.** In a covert mission that not even the Israeli Knesset knows about, Mossad agents infiltrate a top-secret Iranian nuclear site, steal binders of classified information about Iran's nuclear program, and smuggle the information back into Israel so Prime Minister Benjamin "Bibi" Netanyahu can then brag about it on national TV.
- **2020.** President Reuven Rivlin is caught celebrating Passover Seder with one of his children during a nationwide COVID-19 lockdown.
- **2021.** After Israel's attorney general announces his plans to indict Shas Party leader Aryeh Deri on three counts of tax fraud, Deri states, "I thank the Creator of the World for the decision to dismiss all the false accusations against me."
- **2021.** Israeli politicians force the country to hold its fourth national elections in under three years.
- **2022.** Israeli politicians force the country to hold its fifth national elections in under four years.

Trait #4: *Shivyon*

Twenty years ago, when Joel was a young tank soldier in the IDF, he and his army buddy Tomer were off duty, strolling through Tel Aviv, when they saw rock star Shlomo Artzi walk into a café.

Joel was starstruck. "Come on!" he said, tugging Tomer's arm. "Let's get his autograph!"

"Why?" Tomer replied. "Is he better than me?"

This is a perfect example of *shivyon*, Hebrew for "equality." The idea is *not* that Israelis believe everyone is equal. Far from it! Instead, it's that no Israeli wants to think someone else is better than him. Asking Shlomo Artzi for an autograph would have been a tacit acknowledgment that he was somehow superior, which is why Tomer refused to do it.

---

3. "Quotes of the Week," *The Guardian*, October 7, 2007.

Like *dugriyut* and chutzpah, *shivyon* is related to the concept of New Jews. As persecuted minorities in their former countries, Old Jews were constantly aware of their place in the social order (typically the bottom). Deference to the powerful was the norm – and pretty much the only way to survive.

The Zionist pioneers did away with all of that: no more "Yes, sir" and "*Ja, Herr Meister*"; no hierarchies; and no more suits, ties, and other markers of social class. In the land of the New Jew, it was all for one and one for all – *we'll march out into the fields and spread manure together!*

This kind of no-one-is-better-than-me equality is exacerbated by another factor: the small size of the country, both geographically (Israel is the size of New Jersey) and in terms of population (nine million). People in the northern city of Metulla read the same daily newspaper and watch the same evening newscast as their fellow countrymen down in Eilat; school starts on September 1 and ends on June 30 for all Israeli kids; school vacation dates are the same; and the ins and outs of daily life are more or less similar for everyone. "Our small size adds to the idea that we're all in this together and part of the same narrative," says Samuels. "It's also why everyone feels approachable. It used to be that you could literally walk up to any political figure and have a conversation with them – it wasn't considered out of bounds, and there was no fear that something bad would happen. That changed with the assassination of Yitzhak Rabin, but that idea of approachability in Israeli society is still here."

And because the population is so small, nobody is ever more than two degrees of separation away from anyone else. "Gal Gadot, the big-time *Wonder Woman* movie star? Yeah, my cousin did basic training with her in the army."

"The author Etgar Keret? He lives in my sister's building."

"Benjamin Netanyahu? His kid hangs out at the surf club where my sister used to waitress. I met Bibi, and you know what? He's no better than me!"

Another unique aspect of Israel related to *shivyon* is that everyone is accessible, and more so, *willing* to be accessible. Samuels explains,

"Our first prime minister [Ben-Gurion] didn't live in some fortress with security fences. He moved to the desert to live in a cabin! That makes a statement about equality and how available they are to the public."

In Israel, you can reach out to a member of Knesset (Israeli's parliament) on Facebook or get a TV actor's phone number from a friend, and they'll actually respond. No one's unreachable or "out of your league," and if they act like they are, it's a turnoff. It's one of the reasons we were able to interview so many experts for this book: even the country's highest achievers were willing (and eager) to give us time.

In modern Israel, almost all relationships operate on the principle of *shivyon*. Students call teachers and principals by their first names, civilians argue with cops, and soldiers question officers' orders. Many *olim* express frustration over the lack of authority (at least, any that anybody's afraid of). "It's the Wild West," says one veteran immigrant who lives outside Jerusalem. She tells the story of how her next-door neighbor hired a contractor to do home renovations, without the necessary permits. If completed, the addition will block her scenic view of the Sataf Forest. "I called city hall, but they did nothing. I tried the Ministry of Housing – they sent someone who told the neighbor to stop, but he didn't. There are rules, but they have no bite when no one enforces them."

Samuels makes sense of it like this: "This is the Middle East. For thousands of years, rules have been meant to be broken. In a different country with less solidarity between citizens, clear rules and boundaries are needed to keep society functioning. Israelis are bound together by other things, whether it's fighting for survival or thirty-five hundred years of peoplehood. Less weight is placed on the enforcement of rules."

In that sense, the immigrant's story about the home contractor is a metaphor for Israeli society. Because even when the supposed authority attempts to enforce the rules, people don't listen. And why should they? Rules are meant to be broken. These cops, these mayors, these politicians … they're just people, not rock stars.

And even if they *were* rock stars, they're still no better than me!

Trait #5: *Kehilatiyut*

In the Israeliness Quiz, we met the character who's flummoxed that someone would actually *choose* to make aliyah. There's a reason: life here is hard! It's expensive, the government's unstable, Hamas wants to kill us, the schools are a mess, Hezbollah wants to kill us, drivers are reckless, taxes are high, and Islamic Jihad wants to kill us.

So why *would* anyone actually choose to live here?

One huge answer is *kehilatiyut*, Hebrew for "community." Israelis, even complete strangers, share a bond, a palpable connection – the result of our shared narrative and mission to survive and thrive as a people. "In Israel, the ethos of collectivism is strong," says anthropologist Katriel. "The concept of *gibbush* [social cohesion] is cultivated in schools from a young age and continues through the army, into the workplace, and beyond."

That's the magic of Israel. While Israelis won't hesitate to tell you that life here feels harder than in other Western countries, many of them will also tell you that they wouldn't want to live anywhere else. Where else can they feel part of something bigger than themselves, be it the Zionist project or just a country so small that everyone seems to know each other, and everyone's contribution means that much more? Which helps explain why Israel usually ranks near the top ten for the world's happiest countries.

How does *kehilatiyut* manifest in Israeli society? In countless ways. *Tachlis* (bottom line), it comes down to how Israelis relate to one another. The walls between people are much lower in Israel (if they even exist). You can strike up a conversation with anyone – the bus driver, the nurse at the hospital, the guy selling popsicles at the beach. Forget small talk, you'll find yourself sharing personal details. (Returning to the carpool examples from the *"Dugriyut"* section, the inviting communal nature of *kehilatiyut* is what allows for the honest *dugri* speech to come out.) Friendships are deep, and even casual interactions are authentic. Yes, people can be "prickly," but they're *real*. It feels like you're part of a family.

By the way, this bond is something we feel with our Arab neighbors, too. Though Arab-Jewish relations have a long way to go (with that kind of hard-hitting analysis, no wonder we're not writing a political book), both nationally and on a personal level, those who have become acquainted across the divide often report feeling more at home with each other than with people from Western countries (as we note in other chapters in this book). Be it family, soccer, where to get the best hummus, or bonding in the hospital maternity ward, our Middle Eastern values make us more alike and connected than you might imagine.

Kehilatiyut also comes out through good old-fashioned Middle Eastern hospitality. You'll never enter someone's house and not be offered *something*, usually within seconds of stepping through the door. When the repairman (finally) shows up to fix the AC, you'll obviously offer cookies, water, coffee... He'll take you up on it, too, and maybe even chat with you at the kitchen table. When Benji witnessed his first Israeli roommate making coffee for the cable guy, he was floored, having grown up in a country where you treat these people like they're hired help, not coffee dates.

First-time visitors often notice the warmth immediately. The cabbie who shows you pictures of his grandkids when you're stopped at a light. The old woman who flags down your car to ask for a ride to the drugstore on the other side of town. You feel it most, though, in times of conflict, when people put their differences aside and rally together on behalf of soldiers and civilians under attack.

That's another big aspect of *kehilatiyut*: people are genuinely happy to help someone in need. As new immigrants, we often found ourselves the beneficiaries of Israelis' generosity. Complete strangers, hearing our fresh-off-the-boat American accents, invited us to Shabbat dinners and Passover Seders before they even knew our names. When Benji's car got towed, his taxi driver came to the rescue, calling city hall, navigating the situation, and tracking down Benji's car.

But the story that, in our opinion, best exemplifies *kehilatiyut* is that of Michael Levin, the IDF lone soldier from Philadelphia who was killed in the 2006 Lebanon War.

As his mother Harriet tells it:

When we arrived in Israel for Michael's funeral, I was worried we wouldn't have a minyan [a prayer quorum of ten people] for the burial, and that I wouldn't be able to say Kaddish at Michael's grave.

The morning of the funeral, there was a huge traffic jam in Jerusalem. After an hour of trying to enter the cemetery, we saw it was packed with thousands of people crammed shoulder to shoulder. They were even sitting in the trees.

Only when I reached the grave did someone tell me that the thousands of people were all there for Michael. I had told the few people I knew in Israel that we needed a minyan. When they spread the word, the papers and TV news ran the story about this kid from Philadelphia who'd always dreamed of being an IDF paratrooper and was killed in Lebanon. It got picked up on social media. One man attached a loudspeaker to his truck and drove through Jerusalem, announcing that a lone soldier had given his life for the country and deserved to have Kaddish said for him at his grave.

The masses came from all walks of life. Secular and Orthodox, adults and kids, soldiers in uniform from other units. It was hot, mid-August on the Ninth of Av, and many in attendance were fasting. And yet they came. Even though they'd never met Michael.

The beauty of Jewish law is that it's forced Jews to maintain strong communities throughout our history. You can't pray aloud without nine others with you; the shivah ritual guarantees that mourners are never alone, or hungry – they're overwhelmed with kreplach, kugel, kubbeh, and, most importantly, companionship. Judaism doesn't just value community, it demands it.

One of the oldest ideas in Judaism is *"Kol Yisrael arevim zeh la'zeh"* (All of Israel [meaning the Jewish nation] are responsible for one another). In the early 1900s, as Jews began moving to Ottoman-controlled Palestine with the dream of building a Jewish homeland, this idea took on greater significance. Suddenly, teamwork and self-sacrifice became essential. The pioneers treasured camaraderie and shared mission, values that are ingrained in the Israeli DNA today and kept alive by two powerful forces: youth movements and the IDF.

More than a quarter of a million Israeli kids belong to the Tzofim (Scouts), Bnei Akiva, Noam, or another group; youth sports teams often do community service projects, such as delivering meals to local Holocaust survivors and assembling care packages for soldiers. When eighteen-year-olds join the military, they dedicate at least two years of their lives to a cause larger than themselves, putting their lives on the line for the state. Those who don't get drafted for personal or religious reasons often do a year of Sherut Leumi (National Service), volunteering in schools, hospitals, food banks, and centers for at-risk youth.

Jewish people have always believed that it takes a village (or shtetl, as it were). Children as young as seven or eight ride public buses and play in the park unsupervised – not only because daily life here is remarkably safe, but also because parents know that their kids will be looked after by everyone in the vicinity.

The feeling of shared destiny also inspired millions of Israelis in this village to become activists during the 2023 judicial reform crisis. And when the government was slow to respond to the needs of the IDF and home front after October 7th, *kehilatiyut* mobilized the entire nation to collect supplies, send care packages, and raise money for people in need.

As of this writing, over a hundred Israelis are still held hostage in Gaza by Hamas. The national movement to have them returned is called Bring Them Home Now.

"Magical Thinking": Coping in Times of War

When did you last see your neighbor's wife in her nightgown? Not because you're violating the Seventh Commandment, but because that's how she left home? Also, you're together in a dark room, along with her husband and kids, all of you in your underwear.

It's not a drunken Purim party gone awry, just the reality of Israeli life during wartime. Since the 2005 disengagement from Gaza, terrorist groups have launched thousands of missiles at Israeli population centers, sending people scrambling to the nearest bomb shelter or stairwell the moment an *azakah* (alarm or siren) rings out. When couples don't know if they're actually home for the night, safe sex and coitus interruptus take on entirely new meanings.

Visitors are often shocked to discover that daily life in Israel is remarkably secure and that they feel safer here than at home. But every so often, Israelis are reminded that they live in the Middle East, surrounded by such neighbors as Hamas, Hezbollah, and Islamic Jihad – terror groups who have spent years declaring, with zero ambiguity, their intentions to destroy Israel. When this happens, Israelis' lives are upended in immeasurable ways.

"The narrative we tell ourselves is that Israelis are strong, and we know how to carry on," says Jerusalem-based therapist Caryn Green. Indeed, one of the core characteristics of Israeliness is the ability to remain strong in the face of adversity. In his incredible book *Pumpkinflowers*, about Israel's conflict with Lebanon, author Matti Friedman describes how for many years, the military frowned upon the idea of soldiers crying at their comrades' funerals.

But of course, Israelis are only human. The stress they feel during times of conflict does manifest.

When buses explode, when the *azakah* sends families scurrying to the bomb shelter at two a.m., and when parents watch their sons and daughters take up positions along the borders with Lebanon and Gaza in anticipation of the conflict to come, what mental gymnastics do Israelis adopt just to make it through the day?

During the Oslo peace talks of the nineties and Second Intifada in the early years after the turn of the millennium, unprecedented waves of Palestinian bombings rocked buses, nightclubs, and cafés across the nation. Whereas wars were previously waged by armies, suddenly the battles came straight to the home front. "Taking the bus was nerve-racking," says Avital from Tzur Hadassah, just outside Jerusalem.

"I found myself staring at every passenger, judging if anything looked off. If someone wore a bulky jacket in hot weather or looked too nervous, I simply got off the bus. If I stayed, I tried to sit as far as possible from the door, but really, would that even help?"

Avital described her behaviors as typical, resulting in a busload of silent, anxious passengers examining each other up and down. "At that point, terrorists were disguising themselves as ultra-Orthodox Jews, which only increased the tension," she says. "When a Haredi passenger boarded, I'd ask myself, 'Does this check out? Does his beard look real?'"

Aliza Gillman made aliyah with her family in the late nineties and attended high school in the heart of downtown Jerusalem, the site of several bombings. She recalls the arbitrary rules people followed to stay sane. "Many Israelis said, 'We won't go to Jerusalem.' Well, if you lived in Jerusalem, you had to come up with your own rules. 'We won't go downtown,' we said. And if you went downtown, as I often did, you said, 'I won't eat at such-and-such restaurant, because it's always packed, and that's where the terrorist will strike.'"

Though these stories illustrate some of the conscious decisions people make to create routine, much more powerful processes happen beneath our consciousness when living through a wave of terror, according to therapist and Lesley University professor Keren Barzilay-Shechter. When tragedy and stress strike, people rely on defense mechanisms to gain a sense of control; one of the most common coping mechanisms Israelis utilize during intense conflict, she says, is "splitting," through which people divide dilemmas into all-or-nothing extremes. Israelis tell themselves that the 405 bus from Tel Aviv to Jerusalem is either dangerous or safe, or that the next missile is going to fall on their neighborhood or has no chance of doing so.

"Splitting affects how people see one another, too," she says. "We tell ourselves that we couldn't be more different from Palestinians, when the truth is we have much more culturally in common with our 'cousins' (because of our shared forefather Abraham) than with Americans or Brits." (Earlier, we wrote that many Israelis saw Holocaust victims as *freiers* and believed they themselves would have acted differently; this is a prime example of splitting.)

All of these adaptive behaviors are, on the one hand, somewhat logical. After all, if you spot the shooter before he can open fire, you can flee; and if certain cafés or people are, in your mind, dangerous, you can simply avoid them (and hence stay safe).

But as logical as these ideas may sound, they also all contain an element of what the therapist Caryn Green calls "magical thinking" – an irrational belief or superstition that instills a false element of control. The Israeli mother who writes the same text message to her son in the IDF, word for word, emoji for emoji, and at the exact same time every night, as a way to keep him safe... the passenger who, when boarding a bus, climbs the steps right foot then left, one step at a time, and then asks the driver, *"Kol tov?"* (All good?) every time he rides a bus as a way to prevent the bus from exploding...

These and other behaviors that Israelis resort to in tenuous times allow them to keep on living, even if deep down they know these behaviors can't possibly make a difference.

Barzilay-Shechter concurs, adding, "Freud defined defense mechanisms as automatic patterns humans use under a certain level of anxiety. If the anxiety is too hard to deal with, we turn to any means to reduce it."

Unfortunately, as she explains, while they provide a sense of control and, ultimately, psychological relief, these unconscious behaviors also distort reality, consume mental energy, and leave us emotionally drained. "Israelis are a people on *turim gevohim*, literally 'high gears,' or overdrive," Barzilay-Shechter says. And like ticking time bombs, they can be set off by anything. "A terror attack rips open the scars, and we regress immediately. It's a misnomer that Israelis have PTSD. We're not *post*-trauma... we're still experiencing it."

Only as an adult, after habitually breaking into tears in response to fireworks and sudden loud noises, did Aliza Gillman realize she had an undiagnosed trauma disorder originating as early as childhood. In fact, shortly after 2014's Operation Protective Edge, over 40 percent of Israeli children in the border town of Sderot were found to have suffered from anxiety and PTSD.[4]

Considering that the October 7th attacks and war affected people across the entire country, the level of national trauma is even worse. When you also account for the anxiety and depression of hundreds of thousands of displaced citizens, forced to live in hotels, with no end in sight, it is hard to overstate the mental health issues now facing the country.

Even today, Avital says that she often takes the same precautions as in the nineties. "When I travel in New York, I try to avoid being at the center of large crowds. And if I see someone suspicious or can't read a situation, I might change cars or get off the train."

Politics, history, and who did what to whom aside, one reason the Israeli-Palestinian conflict is so maddeningly intractable is that Palestinians suffer from these same phenomena. When millions of ordinary people on both sides have so much unresolved trauma, it's difficult to find empathy for the other and negotiate the actual issues.

Having lived abroad for years, Barzilay-Shechter knows many Israelis become sensitive to their scars only after moving somewhere devoid of the anxiety-inducing triggers commonly found in Israel, like the ubiquitous yelling and constant car horns.

And where you can reasonably expect not to see your neighbor in her pajamas.

Polite versus Nice: What Israelis Say about Us

We could end our study of the Israeli character here, but first let's turn the tables and ask: "How do Israelis see us?"

Word on the street is that Israelis think Americans are overly apologetic, cheery to the point of fake, and hypocritical – the kind of people who'll tell you that it would just be *so great* to get together for coffee,

---

4. Hayah Goldlist-Eichler, "40% of Israeli Children in Gaza Border Town of Sderot Suffer from Anxiety, PTSD," *Jerusalem Post*, July 8, 2015.

but who'll never actually call to set it up, most likely because they had no intention of doing so in the first place. As one sabra friend put it, "Americans are polite but not nice; Israelis are nice but not polite."

Harsh! But again, it's all about cultural norms. For Israelis, the most important thing is to tell the truth, while Americans are more concerned with keeping interactions smooth. Americans are great at inclusion, politeness, and tact, but at the expense of expressing their true feelings and addressing conflict; for Israelis, it's the opposite.

Another Israeli stereotype is that Americans are woefully naive. They lounge around in their mansions (Americans all live in mansions), sipping iced tea (who drinks tea cold?) and spouting elaborate theories on how to solve the Arab-Israeli conflict... Never mind that they've never had to deal with rocket attacks or serve in (and send their children to) the army. When it comes to stereotypes and cultural differences, the "truth" is always somewhere in the middle. For example, neither Joel nor Benji ever lived in even a single mansion. And even though our inner American had plenty of opinions about Israeli policy before we made aliyah, we can also sympathize with the fact that Israelis have only so much tolerance for criticism from those who don't have to run to the bomb shelter with their crying children at three a.m.

This clash in values can lead to a communication breakdown when the two peoples get together – something Hanoch Greenberg knows as well as anyone. Hanoch is the director of the Summer Shlichim (emissaries) Program at the Jewish Agency, where he trains Israelis just out of the army to work at Jewish summer camps in North America. Hanoch also spends every summer at a camp in the southeastern United States. "A big part of my job is to help young Israelis prepare for the culture shock they're about to experience," he says. "To go from a place where people say exactly what they feel to one where they dance around the subject without ever really saying it... it's a big change for Israelis."

This culture shock is one Hanoch often feels himself. He relates this story:

A bunch of us were sitting in the camp office for our daily staff meeting. It was hot, so I shut the window and turned on the AC. A few minutes later, Linda, an American friend I've worked with for ten years, started rubbing her arms, then coughed, then sniffled. "Brrr!" she said a few minutes later.

I looked straight at her and said, "Linda, would you like me to turn off the air conditioner?"

"Yes," she replied.

I got up and shut it off. *"Next time, just ask."*

Here are some examples of how Americans and Israelis say the same thing differently:

The American says...	The Israeli says....
Let me think about it.	I don't like it.
There's a lot of really great stuff in here.	I don't like it.
It was...(dramatic pause to choose words carefully)...interesting.	I don't like it.
It's not that I don't like it...	I don't like it.
I'm almost wondering if we should rethink the whole thing from scratch.	I don't like it.
If I were you, I might...	I don't like it.
It'd be great if you could get it to me before the weekend.	I need it by Friday.
Ideally, you'll have it in my hands by Friday. I mean, it's not the end of the world if not, but let's shoot for that if we can.	I need it by Friday.
Sooner is better than later.	I need it today.
He's a few cards short of a full deck, if you know what I mean.	He's an idiot.
Don't take this the wrong way, but...	???

Our advice: when dealing with Israelis, just say exactly what you think without fear of judgment. They'll appreciate your *dugri* approach – and you'll enjoy the freedom of speaking your mind.

Welcome to Israel!

Face to Face with Cultural Trainer Osnat Lautman

Since 2010, Osnat Lautman has helped thousands of Israeli and international businesspeople and organizations to engage effectively with each other. We spoke with Osnat about some of the difficulties Western clients face when doing business with Israelis and the cultural gaps that both sides need to overcome when working together.

Benji Lovitt: You've made a career of helping Westerners and Israelis bridge the cultural divide. What are the main challenges for Israelis when communicating with their counterparts overseas?

Osnat Lautman: The language barrier, for sure. But more than just the vocabulary itself, it's *how* each side communicates – specifically, the difficulty of translating between direct, honest Israeli speech and indirect, diplomatic Western speech. When an American says they're "a bit disappointed," are they really just a *bit* disappointed, or are they actually quite upset but afraid to say it because they don't want to hurt the other person's feelings?

So how do you teach Israelis to understand what Americans mean?

I always advise Israelis to listen for what I call "downgrading words," like "somewhat," "a bit," or "a little." In many cultures, including the US, people use what's known as "downgraders" to help them save face or avoid hurting or getting hurt. I also tell Israelis to listen for intent to follow up. A phrase like "let's schedule a meeting in two weeks to discuss further" is more meaningful than a non-specific one like "sounds good" or "we'll be in touch."

How about email – does that pose a different kind of challenge?

Definitely. And one of the biggest sources of misunderstanding is simply how many words people from each culture use to make a point. Westerners tend to be more verbose, beginning with a long, formal greeting: "Dear Jane, I hope you had a great weekend and that this email finds you well." Israelis, on the other hand, tend to jump right in, often without a greeting, and have no problem getting an email that just says, "No."

It helps to have a sense of humor when we find ourselves in confusing cultural situations. What do Israelis laugh at when it comes to American corporate communication?

Israelis joke that Americans speak about nothing, and that the only things we Israelis are allowed to discuss with them are football, which we know nothing about, and the weather, which never changes here. This is the complete opposite of how Israelis relate – we connect with each other by talking about our families, upbringing, finances, the hardest moments of our military service, the challenges we face in our work and personal lives, and other topics that Americans consider personal and off-limits. I would argue that talking about these topics builds trust and strengthens the effectiveness of the team as a whole.

How about the reverse: What do Americans find funny about Israeli communication?

That we have no filter! We're loud, we stand close, and Americans think it's absurd that Israelis ask each other what their salaries are and how much they paid for their house – total no-nos in America. I should add that our behavior is more shocking to Americans than theirs is to us, because we're already much more familiar with Americans thanks to movies and television. Although, if I'm being honest – a phrase that goes without saying in Israel – Americans don't find these things funny as much as they're bothered by them.

After some bad service experiences in the States in recent years, I've realized that I am guilty of romanticizing and exaggerating certain generalizations. I've always complained about Israeli customer service, but the truth is it isn't all that great abroad, either, and often it's worse. Do you ever talk about this, how our stereotypes can be wrong?

It's so important to understand the difference between stereotypes and generalizations. A stereotype is to believe that everyone from one group acts in a certain way. Generalizations, meanwhile, help us understand that the majority of a people will behave in a certain way, though not everyone will. Generalizations help us make sense of the world and navigate unfamiliar situations, but we of course need to be careful not to write someone off before we know him or her as an individual.

Is it fair to say that humans aren't nearly as good at categorizing each other as we think we are?

What I've learned in my work is that everyone overestimates how diverse their group appears to others. When I work in Michigan, people say, "I'm from West Michigan – we're different from East Michigan." People on the Upper West Side of Manhattan see themselves differently from those on the Upper East, literally three blocks away. But from the outside, people see them as plain old Israelis, Michiganders, and New Yorkers. Only someone who is local will pick up on the small nuances. Between all the different cultures and levels of religiousness found here in Israel, it's only natural that we'll display a wide range of behaviors and cultural norms. And yet most of these are significant only to us. To the outside world, we're just loud, pushy, way-too-personal Israelis.

Jewish Life in a Jewish State

TV AND RADIO STATIONS THAT GO DARK ON YOM KIPPUR. Child custody laws based on the Talmud. Special *shomer Shabbat* (Sabbath observant) car insurance policies for drivers who promise not to use their cars on the Sabbath.

Strange things happen when Jews get a country of their own. In chapter 2, we examine the intersection of Judaism and daily life. From critical life-cycle issues such as marriage and divorce to even the small details of bus schedules and media programming, Judaism finds its way into just about every aspect of day-to-day existence in Israel. Sometimes the result is oppressive, especially in the eyes of non-Orthodox and non-Jewish citizens. Other times, the result is what many consider the most special part of living here – that finally, after thousands of years of wandering in exile, Jews have a country where we can live according to our own customs and laws.

The Rabbanut: All the Kosher Pork You Can Eat

On November 16, 2018, El Al flight 002 from New York en route to Tel Aviv was forced to make an emergency landing in Athens. The reason wasn't a terrorist attack, a passenger who'd fallen ill, or a mechanical problem, but something far more serious: it was almost Shabbat.

When Theodor Herzl envisioned a Jewish state, did he imagine the world's only *shomer Shabbat* airline? Probably not. But he also likely didn't envision a country with no separation between church – er, synagogue – and state, or anticipate that so much power would be concentrated in the hands of the Rabbanut, the country's Chief Rabbinate and Orthodox rabbinical authority. It oversees all matters of religion, including marriage, conversion, and burial, but its reach extends also to airline flight schedules, car insurance rates, and other aspects of civil life that, at least on the surface, appear unrelated to religion.

The Rabbanut's outsized power is mainly the result of political jockeying and Israel's parliamentary system of government, in which small religious parties wield enormous influence over larger parties that need them in order to stay in power. It's also the cause of one of the strangest ironies in Israel: the justification for a Jewish state was so that after thousands of years of persecution, Jews could worship and live as they please, and yet many Jewish Israelis are affected by official policies regarding religious observance that don't apply to their fellow Jews around the world.

Case in point: if a family wishes to celebrate their child's bar or bat mitzvah at the Kotel Hama'aravi (the main plaza of the Western Wall), the men and women must pray separately, divided by a *mechitzah* (barrier), in conjunction with rules set out by the Chief Rabbinate, with the result being that half the party will have to stand on chairs and peer over the barrier to watch the ceremony. Families who wish to pray together are relegated to a small egalitarian section of the Kotel set aside for non-Orthodox Jews called Ezrat Yisrael; not only can you not physically approach or touch the wall in this section (since a massive stone fell from the wall in 2018 and has yet to be removed), but some families have experienced harassment by ultra-Orthodox extremists who disapprove of Reform and Conservative religious practices.

Another example is transportation: in Israel, almost all public transport shuts down for Shabbat from Friday sundown to Saturday evening. So unlike a Saturday morning in New York, where you'd ride the subway from Queens to Manhattan, transfer to the M31 bus to Penn Station, and then take an Amtrak train to arrive in Boston by afternoon, taking a weekend trip in Israel requires owning a car. You can always rent, of course – but you have to do it (and pay for it!) for two days beginning Friday, since rental car agencies also close on Shabbat. In this way, Israelis are burdened with restrictive travel conditions that aren't imposed anywhere else.

Banks, most supermarkets, and certain retail shops are also legally required to close on Shabbat – or face a hefty fine. Restaurants that stay open on Friday nights or Saturdays are not certified as kosher by the Rabbanut, even if the food they're serving *is*. (This is why you'll find almost no kosher vegan restaurants in Tel Aviv.) And if you want belly dancers to perform at your wedding reception, you're out of luck: the Rabbanut deems the practice immodest, and Orthodox rabbis might refuse to perform the ceremony.

One of the most bothersome aspects of the state's scrupulosity is the lack of civil marriage. "If two people want to marry in Israel, they can't just go to city hall," says Ori Narov, a civil rights lawyer at the nonprofit Religious Action Center in Jerusalem. "Israeli law dictates that they marry in accordance with the religious law of their religion. In the case of a Jewish couple, an Orthodox rabbi must perform the ceremony, and he'll do so only after both bride and groom have proven they're Jewish according to Orthodox Jewish standards. At this time, Reform and Conservative converts to Judaism are not recognized as Jews by the Rabbanut."

Not surprisingly, many secular Israelis find this heavy-handedness both overly intrusive and unfair. "The problem is not that Judaism is our official religion," Narov says. "It's that there's no flexibility. The state forces you to live according to Orthodox law even if you're not Orthodox – or don't believe in God at all."

With that said, the Rabbanut is only in charge of Jewish marriage and divorce. During the Ottoman Empire (from the early 1500s to 1917), the local leadership instated what's known as the millet system, in which each religious group set up its own religious courts. That system is still in place today. In practical terms, it means Muslims marry and divorce according to Islamic law, Christians by Christian law, and so on. Which sounds fair – unless someone from one religion wants to marry a person from another.

"Can't happen," Narov says. "There is no intermarriage in Israel."

While that might make your Jewish mother happy, it's a burden on young Israelis. When a couple applies for a marriage license, the Rabbanut conducts extensive research to verify that both bride and groom are Jewish according to Orthodox standards. If they find that, say, the groom's maternal grandmother is not Jewish, according to Jewish law, neither the groom's mother nor the groom is either, and therefore, they will refuse to issue the marriage license. Situations like this occur most often with immigrants, particularly from the former Soviet Union.

For this reason, many young Israelis these days choose to get married overseas – often in nearby Cyprus – either because one of the parties is not legally Jewish, or to skip the hassle of dealing with the Rabbanut altogether. Upon returning to Israel, bride and groom (or groom and groom, or bride and bride) are considered legally married. Funnily enough, even though the Rabbanut does not recognize non-Orthodox marriages performed in Israel, it *does* recognize marriages performed abroad, including those between same-sex couples.

As a compromise to many people's desire for a less religious society, the government has responded with some pretty strange rules. Tel Aviv now offers limited bus service throughout the city – though rides are free, to prevent passengers from having to break the laws of Shabbat by spending money. Ben Gurion International Airport is open every day of the year except Yom Kippur, and all airlines operate around the clock – except El Al, which, by law, cannot fly on Jewish holidays and Shabbat. And, as of summer 2022, Israeli couples who want to avoid dealing with the Rabbanut can get married in Tel Aviv by getting married in Utah.

No, that's not a typo. During the COVID-19 pandemic, the state of Utah began performing civil marriages online. Israelis who utilize the service can then have their marriages recognized by the Israeli government. (Right-wing religious parties are, not surprisingly, protesting.)

And the strangest compromise of all? That would be the Meat Law, passed by the Knesset in 2019. As the *Times of Israel* reported in an article entitled "Demanding the Impossible,"[1] after years of petitioning from chefs and restaurant owners, the Meat Law finally allowed restaurants to import pork from overseas, with one catch.

The pork must be certified kosher.

Marriage and Divorce: "The Blacklist"

Fifteen hundred years ago, the rabbis of the Talmud decided a woman was unable to marry while pregnant or nursing a child under two.

"The logic is actually beautiful," says Rachel Stomel of the Jerusalem-based Center for Women's Justice (CWJ). "The rabbis feared the new husband would forbid his wife from nursing a child who wasn't his own. Perhaps out of jealousy, or because it could interfere with his ability to get her pregnant. They were trying to safeguard the well-being of mother and baby."

Does it make sense that this kind of law would still be enforced by the State of Israel today? Most Israelis would say no.

But it is – as well as other laws that prohibit women from marrying whom they choose and keep them trapped in marriages they don't want. All because of the power of Israel's Rabbanut.

There is no civil marriage or divorce in Israel; instead, the state imposes religious law on all Israeli citizens, regardless of their own personal religious affiliation: Sharia law for Muslims, canon law for Catholics, and a very narrow interpretation of Orthodox Jewish law for Jews. Those who are unable to marry in accordance with religious rules simply cannot legally marry in Israel, period.

---

1. "Demanding the Impossible, Israel Says Pork Can Only Be Imported If It's Kosher," *Times of Israel*, November 3, 2019.

Let's look at some not-so-obscure scenarios where Israel's marriage and divorce laws can totally screw women over.

Scenario One: The childless widow. Erez and Michal decide to marry. Michal is an only child, Erez the oldest of three boys.

A month after the wedding, tragedy strikes: war erupts, Erez is called for military duty, and he's killed.

"I will never love again!" Michal vows at the funeral.

But life goes on. Three years later, Michal meets Amir, and they decide to marry.

Except they can't. The Torah states that if a husband dies childless, his widow is obligated to marry her late husband's brother. The only way out is if Erez's brother "releases" her through *chalitzah*.

"The *chalitzah* ceremony is still required by Israeli law today," Stomel says. The ceremony is as straightforward as it is bizarre: the widow removes her brother-in-law's shoe and then spits on the ground while witnesses look on.

Particularly troubling is that aside from being humiliating, *chalitzah* leaves women vulnerable to extortion. "We have cases in which the brother demands part or all of his sister-in-law's inheritance in exchange for releasing her," Stomel says. "Or even the deed to her home."

So she must pay up or become an agunah, a "chained" woman who, despite not being married, is ineligible to marry again.

Scenario Two: Widowed with children. Now suppose that Michal did have children with Erez before he died. In this case, Michal would be forbidden to marry her brother-in-law, based on a related biblical law. If she and her brother-in-law were to have children anyway, the kids would be deemed *mamzerim* (plural of *mamzer*, biblical Hebrew for "bastard") and registered on the government's blacklist of *pesulei chitun* (people unfit for marriage, unofficially referred to by some as "the blacklist"). *Mamzerim* are restricted from legally marrying in Israel, as are their descendants.

To be clear, it's not that tons of Israeli widows are clamoring to get married to their late husbands' brothers; the point, Stomel says, is the absurdity. "How is it that in twenty-first-century Israel, the state can either require or forbid a woman to marry the same man, depending only on whether she has children?"

Scenario Three: The agunah. After three years of a tumultuous marriage, Chaya tells her husband, Lavi, that she wants a divorce. Lavi agrees, but only on condition that Chaya pay him a hundred thousand shekels.

Blackmail? Yes. But unless she pays up, she's stuck. According to the state rabbinic courts that oversee marriage and divorce, only the husband can grant the Jewish document required for a divorce known as a *get*. Until she receives a *get*, the woman is an agunah – a "chained" woman.

"Thousands of Israeli women are trapped in marriages because the husband refuses to give the *get*," Stomel says. Some men simply refuse to grant one, out of spite or as a form of control. An estimated one in three women going through divorce experience some form of extortion.

"The Rabbanut polices women's bodies and lives in countless ways," says Dr. Susan Weiss, executive director of the CWJ. "It's not just the women who suffer, but also their children."

As an example, Weiss cites a law concerning divorce and paternity: when a child is born within three hundred days of the mother's divorce, the mother's ex-husband is listed on the birth certificate as the father, regardless of who the actual father is.

"So imagine a married woman has an affair, gets pregnant by her new lover, divorces her husband, and then marries the new guy," Weiss says. "Unless the woman protests, the ex will be listed on the birth certificate as the child's father. If she *does* protest, the child's father will be listed as 'unknown.'"

To actually get the name of the biological father on the birth certificate, according to Weiss, the mother has to file a legal motion – a time-consuming process whose outcome is far from clear. The upshot:

"In many cases, the biological father is not registered as the legal father. In the eyes of the state, he has no legal relation or rights to his own child."

This might sound like a technicality, but there are real consequences. "Suppose the child needs urgent medical care," Weiss says. "Or suppose the child's biological father wants to take the child overseas, maybe to Disneyland, for his bar mitzvah. The mother's ex-husband could stand in the way. He could deny the child permission to get a passport or leave the country, and he has final say on medical treatment, even though he and the child have no biological connection. He can even show up at the child's school and legally take the child home because he's the 'legal parent.' Can you imagine how scary this is for a woman who left an abusive man?"

The twist, according to Weiss, is that although the Rabbanut enforces this "three hundred days" rule in the name of religious law, its roots aren't even Jewish! It actually stems from English common law, which itself was derived from ancient Rome.

"It's crazy," Stomel says. "In Israel, a woman can be a fighter pilot, a high-tech CEO, or even prime minister. But the one thing she can't do is divorce her partner."

Stomel is emphatic that voting for different leaders is not the way to enact reform. "Our government changes with the wind," she says, "and laws passed by one coalition are changed by the next." Instead, she and her CWJ colleagues pursue change in the courts. She cites a recent case in which an agunah sued her husband in civil court for one million shekels, on the grounds that he'd intentionally inflicted emotional distress by refusing to provide a *get* – and won. "Legislation isn't effective," Stomel says. "Precedent is."

When it comes to describing Israel family law, Stomel doesn't mince words. "It's a theocracy," she says. "People recoil when they hear that, but what else would you call it? The state has decided there is only one way to interpret Jewish law – their way. Anyone who doesn't play by their rules is denied access to civil rights."

As a reminder that few things in Israel are black and white, Stomel is not "anti-religion." Far from it. "I myself am religious, as are many of my colleagues," she says. "But part of loving this place is making it the best country possible. A nation that's both Jewish and democratic with basic human rights for all. Isn't that what Zionism is all about?"

Too Jewish, or Not Jewish Enough?

In strong Jewish communities across America, there's a phenomenon called "*shomer Shabbat* little league" – baseball and softball leagues specifically for observant Jews. All activities are held on Sundays and weeknights, allowing participants to pursue their dreams of playing in the major leagues without having to break the Sabbath.

Such special leagues aren't needed in a country where so many aspects of daily life are run by the Orthodox Rabbanut. If there's one country in the world where *shomer Shabbat* baseball leagues are redundant, it's Israel, right?

You would think so.

But Israel is a land of contradictions, and the synagogue-state conundrum is one of the biggest. On the one hand, yes, the Orthodox Rabbanut *does* seem to stick its hands into everybody's personal business. "This country is too Jewish!" many secular Israelis say.

At the same time, many religious Jews feel Israel isn't Jewish enough. In fact, many Israeli towns *do* have *shomer Shabbat* basketball and soccer teams – if they didn't, kids who are *dati* ("religious," or what in America might be called "Modern Orthodox") wouldn't be able to play, since almost all competitive sporting events occur on Shabbat and Jewish holidays.

"When I was a kid, I wouldn't even bother signing up for the big tournaments," says Anat, a former high-level junior tennis player who is Orthodox. (She asked that her full name not be used.) "It's a bit better today – most tournaments start on Thursday, so religious players can play the first rounds. But if those players win their opening matches, they'll eventually have to drop out, because the finals are on Shabbat."

It's the same for other sports. At the youth level, many basketball, soccer, swimming, gymnastics, and judo meets happen on Saturdays. The end result is essentially that for religious kids, sports will never amount to anything more than a hobby.

In addition to limiting the number of Israeli kids who can participate in these activities, there's another consequence: it's virtually impossible for an observant Jew to climb their way up the ladder to the highest echelons of their specialty. At some point, usually in high school, *shomer Shabbat* leagues phase out, leaving observant Jews with a dilemma – play the sport they love, or drop out to honor their religious values.

What do most choose? Watch a professional Israeli sports team, count the number of *kippot* (head coverings), and you'll have your answer.

Today, Anat runs the coaching program at one of the largest athletic centers in the country. She's pitched her colleagues the idea of a *shomer Shabbat* tennis league (which doesn't exist yet, though such teams do for other sports). Their response was unanimous: "Get me religious kids who want to play, and I'll join."

"We can't start a religious league until there are religious kids playing, but until there's a league for them, religious families won't sign up. It's *beitzah v'tarnegol*" (rooster and egg), she laments, citing the Hebrew "which came first" idiom.

What's true for sports also applies to music, theater, and other performing arts, which often schedule concerts and shows on Shabbat.

Amir Mulian, one of the few religious comedians on the Israeli stand-up scene, explains the challenge from the point of view of a performer. "The only way to get good at stand-up is to get up on stage as often as you can. But if you're religious," he continues, "you have fewer opportunities. The clubs book three shows Friday night and another two Saturday night, typically starting before Shabbat ends. That's five shows other comics are doing and you're not. Who's going to progress further?"

Surprisingly, observant Jews even face challenges in the institution

Israelis typically respect most – the IDF. By law, soldiers are guaranteed the protection of all reasonable religious rights. However, every *dati* soldier we spoke to said that, in practice, they need to fight for these rights and even file complaints against officers who are either unaware of these rights or, more often, simply don't care.

"For the first month of basic training, I wasn't given time to pray," says Sergeant Na'ama, a religious combat soldier in the Search and Rescue Brigade. "Eventually, they gave in and let the religious soldiers pray – but only during the time scheduled to clean the bathrooms and bunks, which led to other soldiers picking up the slack and resenting us."

Another common complaint relates to kashrut. By law, all military food and utensils are kosher; kitchens have separate sets of dishes and silverware for milk and meat. But religious soldiers note that other soldiers, even the cooks themselves, flaunt these laws regularly. First Sergeant Yisroel of the Paratroopers recalls the time he showed up for kitchen duty and found that a squad of tank soldiers had turned the kitchen completely upside down. "They were making cheese omelets in the meat frying pan, boiling milk for cappuccinos in a meat pot, and stirring it with a meat spoon. It was complete chaos."

Not knowing which dishes had been tainted, he told his officer that the kitchen would need to be kashered (made kosher again by thorough cleaning and application of heat). "My officer said no way," Yisroel remembers. "I threatened to call the military rabbinate until they caved."

That night, the entire platoon stayed up all hours kashering the kitchen.

Like so many issues, this is about more than just swim meets, prayer times, and milk or meat spoons. It's about what it means to have a Jewish state, what life in a Jewish state looks like, and the kind of country Israel wants to be.

Anat, the religious tennis coach, believes she knows the answer – and she doesn't like it. "Most people don't realize this, but Ben-Gurion's dream was not to have a Jewish country, but an *Israeli* country."

She shakes her head. "Jewish and Israeli," she says. "You might not realize it, but those aren't the same thing."

"If You Will It, *Then* What?"

"We'll know we have a Jewish state when Jewish thieves and prostitutes conduct their business in Hebrew," Israel's first prime minister, David Ben-Gurion, famously said. So much for the doctor/lawyer/dentist jobs that the stereotypical Jewish parents pine for.

Nevertheless, this quote perfectly captures the duality – and inherent contradiction – fundamental to the State of Israel. A country that is, on one hand, unique by virtue of being Jewish, and, on the other, "*k'sha'ar ha'amim*" (a normal country like any other).

Israelis still struggle with the very definition of a Jewish state. From car rental availability on weekends to "how can a vegan café *not* be kosher?" almost every civil issue raises questions in a way that's both special and maddening. Jewish values are so deeply baked into the country's DNA that one could be excused for not imagining that our state could have evolved any differently.

But the Israel we have today wasn't inevitable. In the early 1900s, leading Zionist thinkers offered several competing visions for an eventual Jewish state. Had any of them won out, Israel would be very different today, starting with its name. Can you imagine taking your gap year in Yehudah (Judaea)? Leading a synagogue mission to Ha'am Hayehudi (the Jewish Nation)? Or visiting your cousins in Tzabar (Cactus)? All these options were pitched at various points.

So what were those alternate visions for a Jewish state, and how would Israel (or Cactus Land) look had one of them come to fruition?

Let's turn to Rabbi Amitai Fraiman, director of the z3 Project: Imagining Diaspora-Israel Relations and scholar on Israeli statehood. As Fraiman explains, while there were countless visions, they more or less fell into five categories, each spearheaded by an enthusiastic leader whose name can be found today on street signs in every Israeli town.

Political Zionism (Theodor Herzl). "More than anything, Herzl [born in Hungary] yearned for a homeland that would ensure material, spiritual, and cultural survivability," says Fraiman. "In other words, a country where Jews could live safely as Jews" – a concept Herzl expounded upon in his groundbreaking books *Der Judenstaat* (*The Jewish State*) and *Altneuland* (*Old New Land*), where he details his utopian vision of what a Jewish state might be like.

As Fraiman explains, the Dreyfus Affair, in which French-Jewish military officer Alfred Dreyfus was wrongfully convicted of treason, shook Herzl and convinced him that Jews would never find refuge on the European continent. "So dire was the need for safe haven, Herzl even batted around the idea of a Jewish homeland in Uganda," Fraiman says.

How would life in Herzl's Israel have looked?

"A lot like it does today," says Fraiman. "A nation with its own airline, sports teams made up of Jewish athletes, Jewish cops arresting Jewish criminals. If anything, he might find it a bit too religious, certainly when it comes to marriage, divorce, and the power of the Rabbanut."

Cultural Zionism (Ahad Ha'am). Born Asher Zvi Hersh Ginsburg in Ukraine, Ahad Ha'am (his pen name, meaning "One of Us") envisioned a place that was religiously secular but culturally Jewish. "Ha'am doubted that anyone would grant the Jews a country," says Fraiman. "But he didn't see that as a problem, or at least not the *only* problem. He believed that solving for the material security of the Jewish people was only part of the issue at hand. Equally important was the need to revive Judaism, and for that it would be enough to have a cultural center that would influence the broader Jewish world. In his vision, a significant Jewish population in Eretz Yisrael would serve as a source of culture and inspiration for Jews everywhere else."

Netflix picking up and distributing TV shows such as *Fauda* and *Tehran*, and Jews around the world reading Etgar Keret and Amos Oz are examples of what Ahad Ha'am envisioned, Fraiman says.

Labor Zionism (A.D. Gordon). For Aaron David "A.D." Gordon, a Jewish state meant, first and foremost, connecting physically with the land. "Gordon believed that Jews had lost touch with nature," Fraiman says. "To truly connect with the land, and make it their own, the Jewish people had to work it with their bare hands. This connection to the land would be the way in which Jews could revitalize Judaism and Jewish life." Unlike most immigrants of his time, Gordon was not a socialist, nor was he secular. His motivations were spiritual rather than economic, as he believed that harsh capitalism corrupted relationships between people and made them apathetic to poverty and suffering.

To get a sense of what Gordon's Israel might look like, just imagine the country as one big kibbutz.

Religious Zionism (Rabbi Avraham Yitzhak Kook). "Rabbi Kook [born in modern Latvia] believed in the restoration of Jewish history. He envisioned Israel as a halachic society, operating according to the laws of the Torah and Talmud, complete with a Temple," Fraiman says.

Though some believed that Zionism was a contradiction to Judaism, Kook thought it could strengthen the Jewish people's faith. He believed that two thousand years in exile had dulled Jewish expression and creativity and that by creating a state, Jews would reach their maximum spiritual potential, engaging in a full dialogue with God.

Fraiman points out that Rabbi Kook's vision of a religious Jewish state was inclusive of non-religious Jews, whom he saw as a means to an end. "Kook understood that reaching the utopian manifestation of sovereign Jewish life would take time, and that it would take secular Jews to make the dream possible. So while certainly not ideal, the participation of non-religious Jews was an essential step in the right direction."

If Rabbi Kook were alive today, he would likely be pleased with the role of religion, including the Rabbanut's jurisdiction over public transport, kashrut, and family law. "And honestly," Fraiman says, "he'd want more."

Revisionist Zionism (Ze'ev Jabotinsky). As the father of Revisionist Zionism, the Russian-born Jabotinsky was revising "Ben-Gurion's practical Zionism," Fraiman says. "Unlike the others, who believed in negotiation as a means to ensure a Jewish state, Jabotinsky was a maximalist. He envisioned the most expansive borders possible, from the Mediterranean to the eastern edge of what was then Transjordan."

Jabotinsky believed that to survive, Jews had to be tough; the country would need a strong army, much like the IDF today. "Jabotinsky valued force to defend against an inevitable Arab opposition," Fraiman says. "That said, he wanted and believed in coexistence and in equal rights for all inhabitants of the future state. In his vision, the prime minister would always be Jewish and the vice prime minister, Arab."

For all the debate about competing visions, Fraiman says, we're forgetting one thing: "An actual Jewish country was something our founders could only *imagine*. They would be elated beyond their wildest dreams that a living, breathing, thriving Jewish state actually exists. One with Olympic athletes, filmmakers, Nobel Prize winners, and an army of its own.

"Would Rav Kook love everything about modern Tel Aviv?" Fraiman asks. "Probably not. But he'd be thrilled that more people are studying Torah than ever before in our history."

Kitty Conundrum: A National Cat-astrophe!

If you've been to Israel, you know the country has a serious demographic problem on its hands. At the current birth rates, they may become the country's largest population. It's time to address the issue nobody wants to talk about: the cats.

Yes, *cats* – Israel's answer to the squirrel, if squirrels were less cute and left your garbage dumpster looking like a fraternity house after pledge night. First-timers can't help but notice the shocking ubiquity of street cats everywhere they look – from suburban basketball courts, where they run onto the asphalt as if the coach sent them in to attempt a last-second buzzer-beater, to the hood of your car, the holiest sites in Jerusalem's Old City, and even the bushes on your street (from which they love to jump out and scare the bejeezus out of you).

According to Israel's SPCA (Society for Prevention of Cruelty to

Animals), there are *two million* stray and feral cats in the country. The *Times of Israel* recently reported that Jerusalem has one of the highest concentrations of cats in the world, with nearly two thousand per square kilometer – the equivalent of 160,000 stray cats in Manhattan.[2] Yikes! If these cats all got together and voted in a bloc, they could run the city.

"When I was an exchange student twenty-five years ago, the stray cats looked like strays: sickly, thin, dull coats," says Dr. Danielle Abrahams, a Jaffa-based veterinarian. "Now? I see fat cats with shiny coats. Many are more cared for than house cats."

Why are there so many darn cats, and how did they get here? The prevailing legend is that, like so many other present-day demographic problems in Israel, the answer lies squarely on the shoulders of one group: the British.

When the British Empire took control of the land of Israel from the Ottomans in 1917, there were already plenty of cats in the region. Scientists have traced their genetic makeup back to the African cat, which the Egyptians domesticated.

Apparently, though, there weren't *enough* cats – at least, not enough to combat what the British considered an even bigger problem: rats. So in the 1930s, the British did what you'd expect a bunch of logical Brits to do ... They imported cats to fight the rats.

It worked! But it also created the cat problem Israel faces today. Thanks to warm temperatures year-round, the population exploded. Moderate conditions allowed them to survive through the winter and females to birth up to three litters a year, each containing up to fifteen kittens. According to a Ynet article quoting an animal welfare policy advisor at the Agriculture Ministry,[3] making matters worse is that Israelis, known for their generosity, feed street cats more than people in other countries do.

---

2. Delphine Matthieussent, "Jerusalem Faces Dilemma in Dealing with Hordes of Stray Cats," *Times of Israel*, May 8, 2019.

3. Ilana Curiel, "One Million and Counting: Israel's Street Cat Population Keeps Growing," *Ynetnews,* May 30, 2021.

"In urban areas, many strays are community cats, practically belonging to a courtyard or building," says Abrahams. "They know they'll get food and attention and can go from place to place, being fed multiple times. While the people are acting humanely, it only makes the population problem worse."

Pet owners like to joke about how their cats are plotting to kill them, but jokes aside, what's the big deal? A few cats never hurt anybody, right? A few, maybe not, but a few hundred thousand leads to bigger problems than you would think. Israel's stray cats are unsightly, spread disease, and throw the ecosystem out of whack by hunting birds, snakes, lizards, and other animals that prey on pests that destroy crops.

To solve the problem, the government has debated various scenarios. One option is what's known as trap-neuter-release (TNR), in which cats are caught, brought to a neutering facility, and then let loose – a seemingly humane solution (if you consider the removal of testes humane), though it doesn't address the immediate ecological concerns created by their large numbers.

Another strategy is the creation of fixed feeding stations in the form of small, trough-like containers placed on streets around town, not only to keep the cats alive, but more importantly, to make them easy to collect in order to be neutered – a plan backed by Jerusalem's mayor.

In 2015, this solution was nearly adopted, with 4.5 million shekels of government funding being earmarked for the sterilization of strays. However, Agriculture Minister Uri Ariel refused to use it. Was he afraid to take action due to pressure from the Big Feline lobby? No, Ariel denounced the mass sterilization of street cats because, as an observant Jew, he claimed it went against God's decree to "be fruitful and multiply."

Thus, like so many issues in Israel, controlling the cat population is also a Jewish one. Scholars note that Jews were among the first – if not *the* first – people on earth to show compassion for the well-being of animals and even to adopt laws to address animal cruelty, a topic known in the Talmud as *tza'ar ba'alei chayim*. These include the laws of keeping kosher to ensure that animals experience a minimal amount of pain when being slaughtered and that working animals are provided enough drinking water and even allowed a day of rest on the Sabbath.

Because of the religious angle, it's not surprising that the challenge of tackling the cat population isn't as straightforward as in other places. Furthermore, like so many things, it comes back to politics. "Why must owners vaccinate their dogs annually for rabies? Because it's a national law," explains Abrahams. "We have no centralized management of the cat problem to dictate policy on spaying and neutering. And with government instability, that's not exactly the top priority. If we're lucky, maybe in another ten years."

Until then? We just have to stay paw-sitive. (Sorry.)

By the way, that mayor of Jerusalem who suggested the setting up of feeding stations? His last name is Lion.

Life inside the Jewish Calendar

For many immigrants and visitors to Israel, one of the most meaningful parts of aliyah is living by the Jewish calendar. There's nothing like seeing commercials for *your* holidays ("Sweet deals for a sweet new year!") and how even driving to the Passover Seder in the biggest traffic jam of the year is special, if you can just block out the honking and marvel at the miracle of it all – that after thousands of years in exile, we've finally returned home to sit in gridlock.

Here's how the Jewish calendar plays out in daily life.

Shabbat. The Sabbath, sundown Friday evening to after dark on Saturday night, every week. Traffic is lighter, and public transport shuts down almost completely. Most families, regardless of observance, eat dinner together. Friday is errand day, as families scamper from butcher to bakery to hummus place. A few hours before sundown, the hustle and bustle magically stops, as a calm blankets the nation.

Rosh Hashanah. The Jewish New Year, two days in September or October. A few weeks before, stores begin advertising Rosh Hashanah-themed sales ("Prices sweet as honey!"). On the holiday itself, many towns set up communal shofar-blowing sites. While observant Jews spend the day in synagogue, secular Israelis hike, go camping, or ride the roller coasters at Luna Park. Along with Erev Pesach, the first night brings the most traffic of the year, as people drive to their traditional family dinner.

Yom Kippur. The Day of Atonement, ten days after the first night of Rosh Hashanah. TV and radio stations go off the air, sending secular Jews to Netflix and YouTube, while again, observant Jews spend the day in synagogue. Bikes replace cars, and people dressed in white flood into the street to socialize. The majority of Jewish Israelis fast on Yom Kippur, abstaining from food and drink for twenty-five hours.

Sukkot. Feast of the Tabernacles, a seven-day period beginning five days after Yom Kippur. Kids get two weeks off school, and many offices close, making the holiday a peak time to travel overseas or to Eilat (expect airfares and hotel prices to triple). People head to town squares to buy the Four Species (lulav, myrtle, willow, and etrog), outdoor lighting systems, decorations, and of course, snap-together sukkahs.

Simchat Torah. Rejoicing in the completion and restart of the annual Torah readings, one day after the end of Sukkot. Traffic is rerouted to make way for street parades and dancing with the Torah. The holiday marks the end of the *chagim* (High Holiday season) and the start of *chazlash* (*chazarah l'shigrah* – "back to the routine").

Chanukah. The Festival of Lights, eight days in November–December. You know Chanukah is just around the corner when *sufganiyot* (jelly doughnuts) start appearing on bakery shelves. With kids off from school and many offices closed, Chanukah marks the second peak travel period of the school year. Towns hold communal lightings of the *chanukiah* (Chanukah menorah), and many schools and sports teams do service projects, such as delivering meals to Holocaust survivors or packing boxes for food banks. The most high-profile event is "Festigal," a live song-and-dance, star-studded extravaganza for kids that tours the country. Imagine a musical version of *Snow White and the Seven Dwarfs* but in Hebrew, featuring Ariana Grande and Kevin Hart.

Tu b'Shevat. Jewish Arbor Day, January or February. Many youth movements plant trees, and schools hold Tu b'Shevat Seders, forcing parents to make last-minute runs to the grocery store for whichever fruit their children are assigned to bring to class.

Purim. February or March. Like Chanukah, Purim is announced by the arrival of baked goods, in this case, *oznei Haman* (hamantaschen – the triangular cookies meant to resemble villainous Haman's ears). Kids and teachers dress in costume the week before, according to themes: superhero day, Bible character day, and so forth. Throughout the day, families deliver *mishloach manot* goody baskets to friends, family, and neighbors. Some cities hold parades called *adloyada* ("until he does not know"), named for the commandment to drink so much alcohol that one can't distinguish between heroic Mordechai and evil Haman. Hmm… probably a good day to stay off the roads.

Pesach. Passover, the Holiday of Freedom, seven days in late March– April. Kids get two weeks off, including the week prior, so they can – are you ready for this? – *help their parents clean the house.* A few weeks before Pesach, supermarkets begin getting rid of bread and covering shelves that contain leavened products, stocking up on matzah and traditional Passover foods. The holiday is the third peak travel period; indeed, the largest Seder in the world is at the Chabad of Kathmandu (second place: Chabad of Bangkok; last place: Chabad of the Vatican).

Lag b'Omer. The thirty-third day of the Omer (a countdown period from Pesach to Shavuot), exactly thirty-three days after the first day of Passover. Families, and especially teens in youth movements, build bonfires in honor of the light that legendary Rabbi Shimon Bar Yochai, the first to teach Kabbalah, brought into the world. Religious Jews flock to Meron for mass celebrations and prayer at Bar Yochai's grave.

Shavuot. The Festival of Receiving the Torah, forty-nine days after the first day of Passover. You know it's approaching when bakeries stock up on cheesecakes and traditional dairy products. Like Memorial Day in the US, Shavuot marks the unofficial beginning of the summer season, if you replace the release of comic book movies with blintzes and unbearable humidity. Synagogues hold all-night study sessions called Tikkun Leil Shavuot (lately, secular institutions in Tel Aviv and other cities have been holding them as well).

Tisha b'Av. The Ninth of Av, one day in July or August. To mourn the destruction of the First and Second Temples in 586 BCE and 70 CE, respectively, as well as several other tragedies on this date, communities hold public readings of the Book of Eichah (Lamentations). Religious Jews fast for twenty-five hours, and stores and restaurants close for the evening. Television stations run programming about *sinat chinam* (baseless hatred), the mistreatment of Jews by other Jews for no reason at all – the religious reason given for so much tragedy on this one day.

Tu b'Av. The fifteenth of Av, one day in July or August. A minor holiday, Tu b'Av is the Jewish day of love, whose origins can be traced back to the ancient Temple period. To mark the beginning of the grape harvest, unmarried girls would wear white and dance in the vineyards in what may very well be the original incarnation of JSwipe. Today, Tu b'Av is commemorated much the same way that Valentine's Day is around the world – with candlelight dinners, picnics in the park, and husbands forgetting to buy flowers.

Though not officially on the calendar, the final weeks of August correspond to another important period in Israel: the unofficial countdown to the High Holiday season known as the *chagim* and, with it, the excuse *"Acharei hachagim"* (I'll get to it after the holidays).

Calendar Quirks: Jewish Time

In Israel, nothing is simple. Not even time.

For Israelis, daily life operates according to two calendars – the

January–December Gregorian one that we're all familiar with, and the fifty-eight-hundred-year-old Hebrew calendar that dates the creation of the world at around 3760 BCE.

The result? A progressive, high-tech country that straddles two worlds, ancient and modern at the same ... well ... time.

Here are eight calendar-related phenomena unique to life in Israel:

1. **Double dating.** Secular events such as concerts, work meetings, and travel reservations are scheduled according to the Gregorian calendar. Jewish holidays and important dates in Israeli history, meanwhile, are commemorated according to the Hebrew one. Newspapers list both dates in the masthead, as do government documents.

2. **Save the date (if you know it).** At any given moment, few secular Israelis know the exact Hebrew date. At best, they could probably tell you the month (but only after thinking about it). This makes it hard to plan; many well-known dates in Israeli history, such as the fifteenth of May (the day Israel declared independence), are celebrated according to the Hebrew date (in this case, the fifth of Iyar). Likewise, although technically it's not against the law to get married on the twelfth of the Hebrew month of Cheshvan, some consider it bad taste – that's the day Israel commemorates the anniversary of the assassination of the late prime minister Yitzhak Rabin, a date most Israelis remember as November 4.

3. **One day more.** Two thousand years ago, rabbis announced the arrival of the *chagim* by noting a full or new moon. Because it took a while for news of the moon's relative illumination to travel out of the country, an extra day was tacked on to certain holidays in the Diaspora. To this day, Passover, Sukkot, and Shavuot are one day shorter in Israel than elsewhere. (On a related note: while Israelis conduct one Passover Seder, Diaspora Jews conduct two. Many Israelis don't know this; if you're lucky enough to find one who doesn't, tell him, and then enjoy watching his eyes widen in shock as he exclaims, "*Mah?!*")

4. **Six days a week.** Because Sunday has no religious connotation for Jews, it's a regular day of the week – kids go to school and adults go to work. Aside from some private schools, Shabbat (Saturday) is schoolkids' only day off.

5. **Good night, see you next year.** In Israel, December 31 is known as "Sylvester," per the Catholic tradition of naming each day of the year after a saint (in this case, St. Sylvester, who died on December 31 in the year 335 CE). The Gregorian new year isn't a big deal in Israel – there's no ball drop, no countdown on TV. Adults and kids go to work and school on both days, unless one happens to fall on Shabbat. Recently, however, December 31 has become a party night for younger people, with bars and nightclubs filling up with expats, locals in their twenties and thirties, and soldiers home on break.

6. **Holiday of the Birth.** December 25 is also a regular day. But thanks to globalization and all those annoying Nativity movies, Israelis are becoming more interested in Christmas, known here as Chag Hamolad (the Holiday of the Birth). Some stores even put up trees; driving tours of elaborate Nativity displays in Nazareth, Haifa, and other cities with big Christian populations are also popular.

7. **Fast faster.** Until 2012, daylight savings time in Israel was determined by the Hebrew calendar. It began on the second day of Passover so the Seder the night before wouldn't start too late; it ended the Friday before Yom Kippur, so the holiday could end earlier and, theoretically, make it psychologically easier to fast. These days, daylight savings time begins on the last Friday in March and ends on the final Sunday in October, with no reward for Jewish holidays, much to the dismay of religious political parties who want to preserve the old system.

8. **Before, you know … that.** To denote years before the year 0, instead of saying BC (before Christ), in Israel they say *lifnei hasefirah* (before the counting).

And just so you know …
Israel is one of many countries that writes the date in the style *day/*

month/year. They also tend to write times according to the twenty-four-hour clock (18:30 versus 6:30 p.m.). So if you're planning to visit Israel and text your cousin to please meet you at the airport at 5:30 on 6/7, don't be surprised if he shows up twelve hours early and one month late.

Israel Has Shabbat

In Genesis 2:2–3, the Torah tells us that on the seventh day, God kicked back and took a breather. (Not a direct translation.)

The Jewish people have sanctified the Sabbath ever since, inspiring the poet and Zionist thinker Ahad Ha'am to say that "more than the Jewish people have kept Shabbat, Shabbat has kept the Jewish people."

If you've been to Israel, you know that Shabbat is more than just a day off; it's a feeling, a vibe, an energy in the air. Stores close, public transportation stops, highways and city streets go quiet – you can literally hear the difference.

You can smell it too: one of the first signs that Shabbat is on the way is the scent of freshly baked challah, rugelach, and other delicacies emanating from the bakeries. No wonder that for many visitors, the Israeli Shabbat is what they miss most when they return home.

Whether spent praying at synagogue or chilling on the beach, playing games with your kids in the park or relaxing at an ashram in the Negev Desert, the motivations are similar: to make the Sabbath unique and different from the rest of the week.

Here are some of the special things that happen when an entire country celebrates Shabbat.

Next stop, one floor up ... and then another floor up ... Because the Torah prohibition on starting a fire on Shabbat applies in modern terms to completing an electrical circuit, many residential buildings and hotels provide a "Shabbos elevator" that stops automatically on every floor without pressing any buttons. It's a less exhausting (though not necessarily shorter) option than taking the stairs. In New York City subway terms, "the express train becomes a local."

Siren song. In a tradition that goes back to shofar blowing in the Temple period, many cities around the country sound a siren to indicate the onset of Shabbat.

Holy marketing! Certain stores advertise their dependability by promoting their hours as "24/6."

Roadside goodies. On Fridays, vendors scatter across highways nationwide with handmade signs, inviting you to pull over to buy flowers, *jachnun, malabi,* and other Sabbath treats.

Trivia and *tashbetz*. Israel's major newspapers publish a special *tashbetz* (crossword puzzle) and trivia quiz every Friday. Solving them with the family around the Shabbat dinner table is a popular tradition. The questions revolve around topics such as Israeli current events, history, the Bible, pop culture, and more, and are guaranteed to out dinner guests as either brainy intellectuals or people who get all of their news from social media.

It's Shabbos somewhere. The traditional "Shabbat shalom!" greeting isn't reserved for religious settings like Friday night services or dinner. It's the way that many Israelis tell each other to "have a nice weekend!" And because the workweek ends Thursday and you don't know when you'll next see someone, it's not unheard of for someone to say "Shabbat shalom!" as early as Wednesday night.

Destination: Sabbath town. Before public transportation shuts down Friday evening, the digital displays on the fronts of some buses read "Shabbat shalom!"

Relaxing radio. For over twenty years, popular radio station Galgalatz has changed their format on Friday afternoons to *sof shavuah raguah* (relaxing weekend) mode with soft, laid-back music. It also lists the Shabbat candle lighting times for different cities and reminds listeners of the weekly Torah portion.

Machine guns 'n' roses. On Fridays, public buses and trains are packed with soldiers in uniform heading home for Shabbat, creating the jarring visual of a soldier holding an M-16 assault rifle in one hand and, in the other, a bouquet of flowers for Mom.

"Sir, Shabbat shalom, sir!" Shabbat is a special time even in the army. Soldiers wear *madei alef* – dress uniforms with beret and polished boots – and the entire base lights candles and recites the Kiddush blessings over the wine before a traditional Shabbat dinner. By law, soldiers cannot train on Shabbat, except in situations deemed necessary for national security.

Free parking. Israel's police force is notoriously hands-off – and on Shabbat, all the more so. Most parking-meter regulations are suspended, and even the most egregious violations go unpunished. Next time you're in Israel, take a Friday night stroll through the neighborhood of your choice and note the various unexpected places where cars, taxis, and buses are parked… including right on the sidewalk in front of your apartment building.

Bop to the beat. Friday afternoons, young hipsters and older hippies welcome Shabbat with a drum circle on the Tel Aviv beach, just north of Yafo.

There are countless other ways that weekends in Israel are different from elsewhere, but it boils down to this: every country in the world has Saturday, but only Israel has Shabbat.

Yom Kippur Bike Sale, Get Yours (and) Fast!

Every year on Yom Kippur – the Day of Atonement, the holiest day on the Jewish calendar – Jews worldwide fast, crowd into synagogues, beat their chests to the recital of an alphabetical list of sins they may or may not have committed, sit through hours-long sermons about how to be a better Jew, plead to God that they will be written (and *sealed!*)

in the Book of Life, and flip through the pages of their prayer books to estimate when it will all be over already.

Everywhere except Israel, that is. Here, some do spend the day fasting, pleading, and repenting. But unlike Jews everywhere else, who generally dread Yom Kippur, most Jews in Israel actually – are you ready for this? – *look forward* to it; indeed, many kids count down like Christian kids waiting for Yuletide.

The reason is that in Israel, Yom Kippur looks like a scene out of a fictitious movie called *When the Kids Ruled the Streets* (coming to Hulu this Tishrei). From sunset Yom Kippur Eve to sundown the following day, the streets, boulevards, and highways are flooded with bicycles as Israelis of all ages, but especially children, ride wherever they darn well please. This is possible only because of one of the strongest non-binding laws in Israel: on Yom Kippur, outside of Arab neighborhoods and some remote areas, nobody drives a car.

How did this tradition begin?

Legend has it that until the early seventies, it was common practice for secular Israelis to drive on the Day of Atonement. But on October 6, 1973, everything changed. Israel was caught completely off guard when a coalition of Arab countries, led by Syria and Egypt, chose to launch their surprise attack on Yom Kippur. Traffic was ordered off the road and the country placed under curfew so the IDF could mobilize. Israel nearly lost the war, but didn't – and in deference to the miracle, Israelis stopped driving on Yom Kippur altogether.

"A beautiful story," says Dr. Eyal Naveh, professor emeritus in the department of history at Tel Aviv University, "but unfortunately not true."

According to Naveh, it was traditional for Israelis not to drive on Yom Kippur even before 1973. It wasn't until the eighties, however, that Yom Kippur turned into a day for mass cycling. The reason?

As Naveh, a summa cum laude graduate of Tel Aviv University, recipient of a PhD from the University of California, Berkeley, and one of the leading scholars on modern Israeli history, explains: "I don't know."

He laughs. "The truth is, nobody knows. It's just one of those things that somehow started and now is part of who we are."

A part to relish if you love the outdoors. Indeed, if you could fly over Israel on Yom Kippur (you can't – the airport and airspace are closed), you'd see swarms and swarms of bikes, snaking through streets in much of the country. Kids on tricycles roll down major thoroughfares. Families embark on long-distance rides from one town to another, meeting relatives and friends along the way. Die-hard cyclists, in spandex and often matching jerseys, cruise in a line, Tour de France–style, on the highway from Haifa to Tel Aviv.

The days leading up to Yom Kippur are "a madhouse," according to Gidon Halamish, a manager at the popular Rosen Meets bike store chain. "September is always our best month of the year. I preorder extra bikes and hire as much staff as possible." When asked if there's a line out the door, Halamish laughs. "Line? More like a *shuk*, everyone shouting and jostling to get their hands on a bike before we run out." And once they do? You hope for luck searching on Facebook Marketplace or Yad Shtayim (Second Hand, Israel's answer to Craigslist).

In fact, it's probably the closest Israel comes to American-style holiday sales: "Yom Kippur Bikes – Act Now, Supplies Are Going… *Fast!*"

This Yom Kippur/bike holiday phenomenon is fascinating, for two reasons. First, it illustrates the best of the country's social cohesion. That millions of people who notoriously argue about everything relating to religion and state can agree to honor *this* tradition is a miracle in and of itself.

But more importantly, it shows the evolution of Judaism when its adherents become the majority. In any other country, it would be called "sacrilege" to eat, drink, cycle, and *enjoy yourself* on Yom Kippur. Because when Jews have been a minority throughout history, we have primarily identified and practiced religiously. In synagogue. Attending synagogue on the high holidays is how Jews outside of Israel remind themselves they are Jewish.

But if you want to understand why most sabras identify first as

Israelis and second as Jews, simply visit Israel on Yom Kippur. Don't want to observe religiously? No problem – you're still part of the Jewish nation, a tribe of millions of people who share a language, culture, traditions, and collective memory.

Of course, not everyone endorses all forms of Jewish observance, especially when the holiest day on the Jewish calendar has been turned into an outdoor celebration. In 2012, the religiously observant transportation minister threatened to shut down funding for Tel Aviv's rental bike system unless the service was disconnected on Yom Kippur. They eventually backpedaled. (Sorry.)

Others defend this tradition as a beautiful, unitive activity. To (liberally) paraphrase the great Abraham Joshua Heschel: "We're simply praying with our bikes."

Sukkot: Thinking outside the Booth

Of the 613 mitzvot (commandments) in the Torah, one of the most challenging to fulfill is "to dwell in a sukkah" – the temporary gazebo-like shelter that Jews construct every autumn, to remember how "the Children of Israel dwelled in sukkot when I [God] brought them out of Egypt" (Leviticus 23:43).

It's fair to say the dwelling is more enjoyable now than it was for our desert-wandering ancestors. Today, people eat festive meals inside, hang their kids' artistic decorations up, sing songs, and even sleep under the stars. It is customary to invite guests over, a tradition captured in the 2004 Israeli hit film *Ushpizin* (Aramaic for "guests," referring to the biblical forefathers who are said to visit during the holiday). Friends, family, reconnecting with nature as summer transitions to fall... Sounds like fun, right? It is. Especially if you're not the one who has to build the darn thing.

As if a day of Yom Kippur repentance weren't punishment enough, until recently, building the sukkah involved plywood, a hammer and nails, and the kind of aptitude needed to assemble IKEA furniture. Nowadays, you can buy DIY sukkot kits that snap together LEGO-style in mere minutes, leaving Israelis more time to enjoy the High Holiday season and peruse the seasonal fairs, full of vendors selling sukkot decorations, electrical lighting and space heaters for chilly nights (or fans for not-so-chilly days), and especially the symbolic *arba minim*, the "four species": *etrog* (citron fruit), *lulav* (date palm branch), *hadas* (myrtle), and *aravah* (willow branch).

Ready to build your own? Before you start, just a few rules we should mention. A sukkah needs to meet specifications laid out in Masechet *Sukkot*, a tractate of the Talmud that lists the rules of how to make your sukkah "kosher" (meaning, "it passes muster," not "so you can eat it"). For those behind on their weekly Talmud study, we'll spare you the questions, quips, and debates. Bottom line, the rabbis concluded that to qualify as kosher, a sukkah must fulfill three requirements:

1. **The sukkah must be temporary.** You can't just use the toolshed in your backyard; nor can you leave your sukkah up year-round as storage for your bikes and lawn furniture.
2. **The sukkah must have at least four sides, one of which can be completely open and serve as a doorway.** In other words, it needs to have at least *three* walls, but it can't be triangular.
3. **The roof of the sukkah must be open.** (Read the following in the classic rabbinical singsong voice with hand gestures for emphasis) "*Wha*-ahht...is the meaning of *ohh*-pen?" To paraphrase the Talmud, the roof needs to be open enough that you can see the stars at night, but not so open that in daytime you'd receive more sunshine than shade.

What makes fulfilling the mitzvah of sukkah-building challenging is not just that it's relatively labor-intensive (much more so than, say, the mitzvah of washing one's hands before meals), but that it also requires that you have a plot of land on which to build. Easy enough if you live in a large suburban house with a yard; less easy for the millions of Israelis living in apartment buildings, often without their own parking space.

So how are the citizens of the Holy Land who don't have the requisite space supposed to fulfill the holy commandment?

Enter the *sukkah minimalit* (minimalist sukkah) built directly into the apartment in a way that's legally kosher, maximally space-efficient, and creative in ways our ancestors could never have dreamed.

"From a design perspective, Sukkot has forced us to come up with ingenious solutions," says Daniel Okun, an architect and owner of Okun Architecture in Jerusalem. "It might seem that equipping each apartment with a balcony would be enough, but it's not. To be kosher, each sukkah needs to have a direct view of the sky, so you can't have balconies built directly above one another."

So architects of apartment buildings have to think out of the booth – er, box.

As Okun explains, there are basically two methods of assuring that every resident can build a sukkah. The first is what's known as *mirpesot drugot* (staggered balconies). In these buildings, each apartment has a balcony that's at least partially unobstructed by the one above it. Seen from the side, the balconies look like a staircase ascending along the facade of the building, allowing each resident to build a sukkah with an unobstructed view of the stars.

The second option is the *cheder sukkah* – a room inside each apartment that, come autumn, transforms into a legally kosher sukkah.

"It's a pretty neat piece of engineering," Okun says. In buildings with a *cheder sukkah*, each apartment has a small room that protrudes ever so slightly from the building's facade, frequently a pantry just off the kitchen. What's critical is that the room juts out a few feet beyond the sukkah-room of the apartment above, but less than the one below. "Again, everyone has a clear view of the stars," Okun says. From outside, this, too, looks like a staircase, except that it appears to ascend into the building, ever so slightly deeper with every floor up.

Still, it would seem that the *cheder sukkah*'s inherent design would disqualify it as being "permanent." To get around this (Judaism always finds a way), the roofs of sukkah rooms can be removed, retracting into the ceiling like a garage door. The retractable roof qualifies the entire sukkah room as impermanent and also allows the resident a clear view of the sky.

To see variations of built-in sukkot for yourself, take a stroll through Bnei Brak or older neighborhoods of Jerusalem. Pay attention to the staggered porches and ascending sukkah rooms – things you probably wouldn't notice if you didn't know to look.

If, however, you're planning to buy one of these apartments, be prepared to shell out some extra cash. "Real estate in Israel is already out of control," Okun says. "But a new building in Jerusalem that's sukkah-friendly?" He whistles. "Those are the most expensive on the market."

Unfortunately, not all buildings in Israel are equipped for personal sukkot. In these cases, residents who wish to fulfill the mitzvah may come together and build a community sukkah in the building court-yard or parking lot.

Others simply take matters into their own hands, MacGyver-style.

"Years ago," Okun recalls, "I had a neighbor who removed a window from his apartment, stuck a long wooden plank outside, and built his sukkah there. Totally illegal, of course. So the cops told him if he didn't take it down within seven days, he'd face prosecution. Which was perfect, because that's how long Sukkot is – seven days."

Israel – where even the criminals have a *Yiddishe kop*.

Yom Hazikaron: DJ in Tears

Every spring, on the fourth of Iyar – twelve days after the conclusion of Passover – a siren sounds at eight p.m., and Israel comes to a complete stop.

The siren, *hatzfirah*, marks the beginning of the saddest day on the Israeli calendar: Yom Hazikaron, the Day of Remembrance for the Fallen Soldiers of Israel.

Unlike Memorial Day in the US, Yom Hazikaron is somber. No barbecues, no parades, no doorbuster sales ("All mattresses 30 percent off!"). Instead, it's a day of ceremonies, graveside visits, and tears.

Lots of tears.

Because most Israelis serve, know people who serve, and raise children knowing they'll serve, the army plays a monumental role in Israelis' lives and consciousness. An American can go a lifetime without ever having a conversation with a soldier; in Israel, that's simply not possible.

It only follows that everyone knows someone – a relative, the child of a work colleague, a friend of a friend – whose life was taken too soon. When a soldier dies in Israel, it's front-page news, always. Stroll through any town and you'll see playgrounds, public parks, and schools named after fallen nineteen- and twenty-year-olds who grew up there. The nightmare of every Israeli parent is answering an unheralded knock on the door and coming face to face with a grim military officer sent to deliver the most unbearable of news.

For these reasons and the sheer number of people whose lives are affected by tragedy, Yom Hazikaron is widely considered the most solemn day for the Israeli people. Sadder even than the most somber day on the Jewish calendar, the Ninth of Av.

Ido Benvenisti has devoted his life to education about Israel, first as a youth movement *shaliach* (emissary) in the United States, and currently as part of the management team for the Jewish Agency's Partnership2Gether network. "My family isn't religious, so I didn't grow up with the Ninth of Av," Benvenisti says. "But everyone is affected by Yom Hazikaron. Even little kids know to stand quietly during the siren."

Like all Jewish holidays, Yom Hazikaron begins in the evening. Before sunset, commercial activity shuts down. Most people either attend or watch a *tekes* (ceremony) on television that begins just after eight p.m. Why after? Because eight o'clock is when the *tzfirah* sounds. For one minute, everyone stops what they're doing and stands in absolute silence to remember and recognize all who were killed during military service.

The gravity of the day is inescapable. Television programming is dedicated to remembering soldiers and their stories, filled with sad testimonials and interviews with their families and loved ones. At ceremonies, people recite the Natan Alterman poem "Magash Hakesef" (The Silver Platter), which quotes the future president Chaim Weizmann's looming words from December 1947: "No state will be given on a silver platter."

The poem describes a pair of soldiers: dressed in battle gear, endlessly fatigued and weary from war, yet so young that they "drip Hebrew dew" of their youth. When a nation in tears asks who they are, the soldiers softly answer, "We are the silver platter on which the Jews' state was presented today."[4] Alterman's poetic and tragic words conveyed that if a state were to be given on a silver platter, it would be represented by the young men and women who would lose their lives defending the country.

Radio stations play the melancholy songs that Israelis know so well, such as "Ein Li Eretz Acheret" (I Have No Other Land) and "Ma Avarech" (What Will I Bless?).

"The music is incredibly powerful," says Benvenisti. "These are songs that you don't typically hear throughout the year, and suddenly they're coming one after another. I don't even need to watch TV to be moved – the songs alone do it."

The sadness is felt by all. In a recent Yom Hazikaron broadcast, the DJ of the popular morning show *Medinah b'Derech* (A Nation on the Go) broke down crying on air as she remembered a close childhood friend who, as a paratrooper, was killed during an operation in Gaza. For several moments the radio was silent, except for her audible sobs.

In some towns, street signs are covered with a black cloth bearing the name of a local resident whose life was lost. High school students pair up with local families who lost a child and accompany the family to the cemetery. At eleven a.m., the nation stops again, this time for

---

4. Natan Alterman, "Magash Hakesef," The Seventh Column, *Davar*, December 19, 1947, translated by the authors.

a two-minute siren. Cars pull over to the side of the road, drivers stepping out to stand.

In 1997, the Knesset elected to expand Yom Hazikaron to honor victims of terror attacks as well. Most Israelis, however, still think of it as a day for remembering soldiers. And while Israel is certainly not the only country to dedicate a day to its fallen heroes, what makes this day unique is its juxtaposition to Yom Ha'atzmaut, Independence Day.

In the first years following its establishment, Israel held ceremonies to mourn fallen soldiers on Independence Day itself. When families asked that they be moved to a separate day, Prime Minister Ben-Gurion established Yom Hazikaron on the day before. Non-Israelis could be excused for finding a day of mourning immediately followed by a day of euphoria simply bizarre. And it is. Heartbreaking memorials leading to fireworks and flyovers...

And yet there is something incredibly Israeli about the juxtaposition, the poignant mixture of sadness and joy. That we cannot reach the emotional high of Yom Ha'atzmaut without first experiencing the low prevents anyone from missing the point: we could never have achieved the dream articulated in "Hatikvah" without the brave men and women of "The Silver Platter" who made it happen.

"It's easy to be cynical, but there's a lot to celebrate," Benvenisti says. "It's just more meaningful because we know the cost."

CHAPTER 3

The Hebrew Language

IN THE 1970S, THE WEEKLY NEWS MAGAZINE *HA'OLAM HAZEH* (This World) published a new word, previously unknown to even the most fluent Hebrew speakers: *l'hizdangef*, meaning "to stroll down Dizengoff Street," the trendy thoroughfare that runs through Central Tel Aviv. Though the word has long since left colloquial vocabulary, the Hebrew language's system of conjugation allows for the constant creation of new verbs to match the times.

Language is a crucial ingredient to Israeli identity. It's been said that what makes the Jews a people is their shared history, their connection to the land of Israel, and, yes, their language. For thousands of years, Jews prayed facing east, yearning to return to their ancestral home. Scattered around the world, we never forgot our history, passing on our values, stories, and traditions *mi'dor l'dor* (from one generation to the next).

But from Poland to Iraq, Russia to Australia, none of our Jewish ancestors dreamed that one day, future generations would establish the first Jewish state since King David – and actually speak the ancient language of Hebrew there.

If the revival of a two thousand-year-old nation sounds absurd, the resurrection of a dead language lags not far behind. To be fair, Hebrew wasn't completely dead, just reserved for religious contexts like prayer, reading from the Torah, and singing the table of contents of the

Passover Haggadah. Not so helpful when you need a suave opening line on a Saturday night.

In chapter 3, we'll explore what Israeli singer Ehud Banai calls "the language of the Hebrew man," from its modern roots to its evolution as a living, breathing language; and how, were it not for the vision of one nutty lexicographer just over a century ago, single Israelis might be trying to pick each other up in Yiddish.

"Nice *shayna punim*, come here often?"

Eliezer Ben-Yehudah: Hebrew Resurrector

Two thousand years ago, Latin was thriving. Today, other than in SAT prep courses and on American coins, it's dead.

If the idea of modern-day Italians reviving Latin sounds absurd, we're with you. Who in his right mind would try to bring back a funny-sounding language that hasn't been spoken since the fall of the Roman Empire?

Eliezer Ben-Yehudah, that's who!

Eliezer Perlman (like so many of Israel's important early figures, he would later Hebraize his name) was a Russian-born lexicographer who dreamed of breathing life back into ancient Hebrew. Like a mad scientist of linguistics, Ben-Yehudah extracted words from the Bible, modernized them according to the needs of daily life, and created grammatical rules. The result is the language that Israelis, overenthusiastic tourists, and a handful of dedicated summer camp counselors speak today.

Now, Hebrew hadn't technically bitten the dust when Ben-Yehudah arrived on the scene; it was still used regularly in Jews' communications with God – when they prayed for a return to Zion, for example. Nor was this revival the first time Hebrew became their official spoken language. Starting three thousand years ago, when the Jews' forty-year journey through the desert finally ended in the land of Israel, Hebrew was the national language for more than a millennium, until the Bar Kochba war in 135 CE. "Like a linguistic Sleeping Beauty, Hebrew then fell into a deep sleep, only to awaken seventeen hundred years later," says Dr. Gabriel Birnbaum of the Academy of Hebrew Language. "Not from the kiss of a prince, but from the determination and vision of one of the most underappreciated, brilliant, and eccentric minds in Jewish history."

Born in modern Belarus in 1858, Eliezer Ben-Yehudah grew up in a religious family. He learned Hebrew in a yeshiva and achieved fluency in a number of languages, including French, German, and Russian. It was in Europe, during what's known as the Haskalah (the Jewish enlightenment), that Ben-Yehudah first encountered Hebrew in a non-religious manner by reading Zionist texts. As Birnbaum explains, Ben-Yehudah quickly became convinced that the Zionist dream of uniting the world's Jews in Israel could only succeed with the revival of Hebrew as a modern, secular language. To survive, Ben-Yehudah believed, the Jewish people, like any people, needed a language to unite them – so he took it upon himself to single-handedly make it happen.

Ben-Yehudah moved to Palestine in 1881 with his work cut out for him. His bold idea had plenty of detractors, foremost among them Theodor Herzl himself. After their first meeting, Herzl dismissed the idea as absurd – and can you blame him? "It may be the language of the Jewish people," Herzl supposedly said, "but who among us can buy a train ticket in Hebrew?" Herzl, who didn't speak Hebrew himself, favored German (regarded back then as the language of high culture) or the Swiss model, in which several official languages coexist.

Ben-Yehudah's biggest critics, though, were the group of people who already had a strong relationship with the Hebrew language: the ultra-Orthodox in the land of Israel. They regarded Hebrew as *lashon kodesh*, the holy language reserved only for prayer. Despite Ben-Yehudah's best efforts to gain their support, they met him with opposition, even attempting to murder him once as he rode a donkey through Jerusalem. To this day, many Haredi communities in Bnei Brak and Jerusalem still communicate in Yiddish.

Even the death threat did not dissuade Ben-Yehudah. He began by creating as many modern words as possible based on roots found in the Bible. For example, to form the modern word for "police" (*mishtarah*), he took the word *shoter* – ancient Hebrew for "government official," which first appears in the fifth chapter of Exodus when Pharaoh instructs his *shotrim* (plural) to make the Jewish slaves work even harder. He took words from other Semitic languages, too, particularly Arabic. When he needed a word for which there was no precedent, Ben-Yehuda did what all great Israeli minds have done – he improvised. One example of this is the word for doll, *boobah* (a reconfiguration of the French *poupée*).

To help spread the new language, Ben-Yehudah began putting out a newspaper. Every issue of *Hatzvi* contained new Hebrew words. By the turn of the century, most Jews in Palestine could both read and understand the language. And though no one had spoken Hebrew as a mother tongue in almost two thousand years, this would change through Ben-Yehudah's secret weapon: children.

He began with his own, pledging with his first wife, Devora, to speak only Hebrew to each other and isolating his son, Ben-Zion, from hearing any other languages. It worked. Ben-Zion Ben-Yehudah was the first Jew in two thousand years whose native language was Hebrew. Like so many legends in their fields, Ben-Yehudah was steadfast in his commitment to his cause; during his wife's final battle with tuberculosis, he smuggled his mother out of Russia but refused to let her speak a word of Russian to his children.

Ben-Yehudah knew that if he could teach children the new language, they would teach their parents (a phenomenon that still occurs with immigrants today). In the late 1880s, pupils in the newly founded settlement of Rishon L'Tzion (First to Zion) began learning through the method of Ivrit b'Ivrit, learning Hebrew via Hebrew immersion. These children weren't just Ben-Yehudah's first students, they were also his guinea pigs – Ben-Yehudah would sometimes come to class with options for new words, and whichever the children liked best, he'd stick with.

"It's quite likely that were it not for Ben-Yehudah, we wouldn't have a functioning Jewish state today," Birnbaum says. "What we were able to create would not have been possible with so many different languages. How would so many diverse immigrant groups communicate?" Birnbaum laughs. "Today we have one language, and it *still* doesn't work so well!"

To understand why Ben-Yehudah is so important, Birnbaum says, we need to appreciate that it wasn't just vocabulary and grammatical rules he gave us, but something more: belief.

As historian Cecil Roth puts it: "Before Ben-Yehudah, Jews could speak Hebrew. After Ben-Yehudah, they did."

The Academy of Hebrew Language

If you meant to go to sleep at ten p.m., but instead you stayed up till three in the morning rewatching an entire season of *Shtisel*, how would you describe in Hebrew what you did?

The Akademiah, Israel's Academy of Hebrew Language and the country's official institution for Hebrew scholarship since 1953, would prefer you say that you engaged in *tzfiyat retzef* (continuous viewing). But no Israeli actually says that; instead, they'd describe what they did as *beenj*, from the English word *binge*.

"Language is a living, breathing entity, and Hebrew is no different," says Dr. Gabriel Birnbaum, an acclaimed linguist and senior researcher at the Academy. Like any living organism, language continues to develop and needs constant updating – especially this particular Bible-based language that lay dormant for two thousand years. Using terminology from the Torah, Eliezer Ben-Yehudah came up with words for modern phenomena like greenhouse, policeman, and newspaper in the late 1800s. But what about "ethernet cable"? "Cryptocurrency"? "Transgender," "lactose intolerant," and "vegan"?

That's where the Academy of Hebrew Language comes in. In a country famous for its high-tech research and development, the Academy, located on the Jerusalem campus of Hebrew University, conducts R&D on something that predates cybersecurity, pharmaceuticals, and, as far as we can tell, vegans: the Hebrew language. They publish dictionaries in the areas of electronics, chemistry, molecular biology, and psychology. They also develop words to describe what happens in the fields of banking, law, and artificial intelligence.

What makes the Academy the source of fascination for many everyday Israelis, however, is their influence on the more commonly used vernacular.

According to Birnbaum, Academy members meet five or six times a year to discuss and debate linguistic issues and new words submitted for approval. The Academy's plenum is composed of linguists, writers, and poets; among them, as in any ideological debate, there exist both conservative and liberal viewpoints. "Purists favor the use of existing Hebrew words whenever possible," Birnbaum says. "They feel it's the best way to strengthen the language's status and keep Hebrew alive." *Tzfiyat retzef* (continuous viewing) is a perfect example.

Progressives, meanwhile, are more open to what Birnbaum calls "loanwords" – terminology and phrases borrowed from other languages, on the premise that they help connect Hebrew speakers to global culture. "Progressives would argue that ancient Hebrew itself was full of hundreds of loanwords from neighboring countries," he says. As an example, he cites the word *Sanhedrin*, the name of the supreme religious body of ancient Judea, which comes from the Greek word *synhedrion* (assembly).

While no hard rule exists, Academy members base their decision on a few factors, including how accepted the term is already in everyday speech; whether the term's *shoresh* (root) can generate verbs and adjectives; and whether a Hebrew alternative would be convenient, catchy, and appropriate. Often, the Academy chooses a middle ground, concerning themselves with the preservation of grammar and syntax of the formal language used in lectures, books, and newspapers while acknowledging that Israelis on the street will use loanwords and slang.

For the most part, the Academy's suggestions are accepted without notice. But every so often, they coin a word that fails to catch on. It happened when Israelis rejected the phrase *tzfiyat retzef* in favor of *beenj*; another example is the word for the internet itself. Ask your waitress for the password to the *mirshetet* (the Academy's suggestion, based on the biblical word that means "net"), and she'll give you a confused look, like Moses staring at a router; but if you inquire about the *eenternet*, or *whyfy*, she'll say, "Zero through eight" (the pretty much universal Wi-Fi password in every restaurant in the country – so much for Israeli cybersecurity).

Unlike many government institutions in Israel, the Academy of Hebrew Language is well loved. One reason is that they actively and enthusiastically engage with the public. Through their social media presence, the Academy crowdsources suggestions for new words and receives thousands of responses from their more than half a million followers. If you have a suggestion for a new Hebrew word, you can email them directly, and they'll consider it. One recent suggestion that passed was *hesket* (podcast), based on the Hebrew root that means "hear" or "listen." (Though most Israelis use the loanword *pode-kahst* for what they listen to while stuck in traffic.)

The other reason that the Academy is so popular is that their work is deeply emblematic of what just may be the biggest contradiction of the Jewish state: an ancient people living in a modern land. Whether they're aware of it or not, the work of the Academy gets to the heart of the most pressing Israeli question of all: Who *are* we as a people? How do we negotiate the balance between who we were thousands of years ago and who we are today?

This question is one the Academy itself faces personally. In Israel, the Academy is known simply as Ha'akademiah, a loanword derived from Greek; as the Academy of Hebrew Language ironically acknowledges on their website, there's no Hebrew word for "academy" – at least, not yet.

Any suggestions?

More than Words: Nonverbal Communication

It's been estimated that 90 percent of interpersonal communication is nonverbal. If you're like most people, no one ever told you this. (Or maybe they did but, come on, who pays attention to *words*?)

Israelis, meanwhile, have known this all along. Like their Mediterranean brethren, sabras are notorious for energetic gestures, grunts, and gesticulations that communicate much while technically saying nothing at all.

Here are eight popular nonverbal sounds and gestures you should learn if you want to communicate like a true Israeli:

1. **Finger Flap over the Shoulder: "Long ago, in a galaxy far, far away..."** To indicate that something happened anywhere between "a while ago" and "a really, *really* long time ago," raise either hand to shoulder height, palm facing you and fingers together; then rapidly curl your fingers down and up as if motioning "come here." Like all great Israeli gestures, this over-the-shoulder finger flap communicates not only information but the agent's point of view. A high school history teacher would *not* use this gesture to indicate when King Solomon began work on the Second Temple; she would, however, use it if asked when she last got a raise.

2. **Claw to the Neck: "They just won't leave me alone!"** To describe a person who incessantly nags, compare them to a blood-sucking *kartziah* (tick) that leeches onto your skin and refuses to let go. Curl your hand into a claw, fingers spread, and sink fingertips into your neck. If you're really annoyed, add a guttural "*Kchhh!*" sound.

3. **Total Body Shrug: "What can I do?" or "Who knows?"** When complete frustration leads to total resignation, the best defense is to accept the utter futility of the situation. Raise your hands to either side, palms up, as you shrug your shoulders, while simultaneously raising your eyebrows and tilting back your head: "Whaddya gonna do?" The more exaggerated the motion, the less hope conveyed that you could possibly do a darn thing. To see this in action, ask

a nearby Israeli why the country recently had five elections in the span of three and a half years.

4. **Flip over the Grapefruit: "What are you thinking?"** When someone has confused or wronged you, and you're wondering just what the heck got into him, show him how stupid you think he is by raising either hand to chest height, palm down and fingers curled like you're holding a grapefruit, while simultaneously rotating your wrist outward, away from your body, so that the invisible grapefruit is now facing the sky. Now add a look of utter disbelief-plus-disappointment, and voilà – the person you're talking to now knows that he's a complete idiot. To see this in action, ride in the passenger seat of a car while a sabra drives; the next time she's cut off in traffic or a pedestrian absentmindedly steps into the street, she'll flip it.

  (Incidentally, this is how left-handed soldiers in the IDF are taught how to pull the safety pin out of a grenade before throwing it. Righties are taught to turn their wrist toward them as if checking the time.)

5. **Invisible Yellow Card: "Patience!"** The next time your kid starts talking to you while you're on the phone, the guy behind you in the checkout line at the grocery store tells you to hurry up, or your boss starts firing off a slew of assignments, tell them to calm the heck down and hold their horses like an Israeli would: Raise your hand to chin height, palm up and with fingers closed, like you're a soccer referee presenting Pelé with a yellow card, and then shake your hand up and down. This gesture is lacking in Western communication except for the occasionally used, less universal, and far more polite "One moment, please!" index finger. Despite its utility, some non-Israelis may find this gesture rude when used forcefully – it's kind of like being yelled at in sign language. This gesture is often accompanied by the word "*Rega!!*" which may or may not be screamed in your general direction.

6. **Tongue Click: "Nope!"** When a simple "no" isn't enough because "yes" would be unthinkable and the person who just asked you

the question should have known better, there's nothing like an aggressive tongue click to put him in his place. The click sounds exactly like the first articulation of the Western "tut-tut-tut" when scolding a naughty child. To hear it, go to a café and then, when the bill comes, ask your server if tip is included.

7. **The Keep-On-Going Karate Chop: "Straight ahead till the end of the world!"** When asked for directions, Israelis are biologically programmed to try to help you, even if they don't actually know for certain how to get there. At the very least, they will send you in the correct direction with a repetitive karate-chop motion that means *"Yashar, yashar, yashar"* (Straight, straight, then more straight!).

8. **By the White of My Eye: "Do you think I was born yesterday?"** When someone says something so ridiculous that it warrants a response of snarky disbelief, pull down on your lower eyelid with your index finger, exposing the bottom hemisphere of your eyeball. "Really? Do you take me for a complete idiot who was born yesterday?" your lower eye says.

Between these and other gestures, grunts, and facial tics, it's theoretically possible for two Israelis to have an entire conversation without uttering a single word. If that makes you think, "Wow, pretty impressive!" think again: The correct response is to simply tilt back your head and go, *"Psshhhh!"*

That's Israeli for "You go, girl!"

Hebrew Is Magic

Why do wars happen?

What's the nature of truth?

What does inner peace look like?

Philosophers, scholars, and ordinary human beings have been debating these questions since the beginning of time.

Could it be that the answers have been ingrained in Hebrew all along?

"The Hebrew language is magical," Dr. Gabi Shetrit, professor of a course entitled "Bible as Literature," tells me in his Haifa University office. Shetrit is *dati*, an observant Jew. But when he reads and teaches the Bible, he does so analytically, as an English professor might dissect Shakespeare. He pays attention to elements that most of us would never pick up on, like alliteration, metaphor, repetitions of words and letters, and even the appearance (or notable lack thereof) of individual letters themselves.

Small and wiry, about sixty years old, Shetrit is the kind of guy who is never not smiling. But when he talks about the Hebrew language, his smile grows especially wide. "In Hebrew, every piece of the word matters," he says. "Words are more than just a collection of syllables and sounds. They are lessons, packed with messages and clues about values, human nature, and how we are supposed to live."

When analyzing words, scholars like Shetrit consider everything from spelling to the parts of the mouth involved with pronunciation: lips, teeth, tongue, roof of the mouth, and throat. But the greatest insights about Hebrew come from the *shoresh* – the three-letter "root" from which the word is derived. As it turns out, a deeper look beneath the root of a modern, mundane word can unlock thousands of years of deep meaning and philosophy.

Here are some of Shetrit's favorite hidden life lessons embedded in everyday Hebrew vocabulary.

Time. "The word for time, *zman*, is actually a combination of two shorter words," Shetrit says. "*Zeh*, which means 'this,' and *man*, biblical Hebrew for 'what is.'" This, as Hebrew instructs, is how we should conceive of time: as simply what is, in the present moment, now. Eckhart Tolle has sold more than three million books based on this idea, but Hebrew knew it first.

Opportunity. Shetrit explains that the Hebrew word *hizdamnut* contains three other words within: *zman* (time), *matanah* (gift), and *todah* (gratitude). The lesson? "See every opportunity as a gift, be grateful for it, and pursue that opportunity now, because it won't last forever."

Morning. Scientists discourage checking our smartphones immediately upon waking up. Why? Because this is when our minds are most fertile, creative, and ready to learn. "But Hebrew already knew this!" Shetrit says enthusiastically. "The Hebrew word for inspection, *bikoret*, is derived from the word *boker*, which means 'morning.'" If you want to be more productive, do important things in the morning, when your mind is readiest to investigate!

Ignorance. Why does it hurt to be ignored? The word *l'hitalem* (to ignore) contains the answer. "We think of ignoring as something we do to someone else, but *l'hitalem* is actually a reflexive verb, one where the subject takes action on itself," Shetrit explains. "Like *l'hitlabesh*, to dress oneself, or *l'hitpaer*, which means to put on makeup." So what, exactly, are we doing to ourselves when we ignore someone? The key to the entire puzzle is the root *olam*, which means "world": when we ignore someone, we build a world for ourselves and only ourselves, one in which the other person is not welcome. This, Shetrit says, is why it hurts so much to be ignored. As human beings, what we want more than anything is to belong; when you're ignored, the other person constructs a world that does not include you.

Truth and lies. The word for truth, *emet*, consists of three letters: *alef*, the first letter of the Hebrew alphabet; *mem*, the middle letter; and *taf*, the final letter. *Sheker* (lie), meanwhile, consists of three consecutive letters that are rearranged (like Q-R-S becoming S-Q-R). According to Shetrit, the difference between truth and lies is not the presence or absence of facts, but how they are presented. "The truth encompasses all facts, from beginning to middle to end," he says. "A liar, however, selects a mere portion of the facts, rearranges them, and presents them as the entire story." This also explains why liars are so dangerous: their stories contain just enough of the truth to be believable, while leaving out the parts that don't serve them.

Gossip. Ponder this: the word *rechilut* (gossip) is derived from the same root as *rochel* (peddler). Why might that be? "The person who engages in gossip is a salesman, too – just a different kind," Shetrit says. "Just as a carpet salesman sustains himself by selling carpets, the gossiper sustains himself by going door to door spreading lies. His lies are essential to his livelihood."

Marriage. Hollywood depicts marriage as passionate kisses and midnight strolls along the sea. Hebrew paints a more honest picture: the word *nisu'in* (marriage) derives from the root *naso*, which means "to carry something heavy." It might not sound romantic, Shetrit says, but those who've been married a long time know it's true. "Marriage is an effortful journey by two people working together to carry something forward – something precious, but also heavy."

War. *Milchamah* (war) comes from a word that even beginner Hebrew-speakers are familiar with: *lechem* (bread). "All human conflict is, ultimately, about the struggle for bread," Shetrit says.

Peace. One of the first words children learn in Hebrew school is *shalom* – Hebrew for "peace," and also used as "hello" and "goodbye." But where does *shalom* come from? The root, Shetrit explains, is *shalem*, "whole" or "having a sense of being complete." It's an important lesson, he says – not just for the world, but for ourselves. "Hebrew tells us, loud and clear, that peace is not about the absence of conflict, but about a sense of feeling complete with whatever the situation happens to be." The key to inner peace is not to forget or deny conflict and traumas, but to accept them as part of our completeness.

From the Hellenists to the Romans and the Mamluks to the Ottomans, Jerusalem has changed hands numerous times over thousands of years. Ironically, despite centuries of fighting, the holy city's Hebrew name Yerushalayim contains the same root as the word "peace," *shalom*. However intractable the fighting may sometimes appear, let us hope that, once again, perhaps Hebrew knows something about our future that we do not.

Face to Face with Liron Lavi Turkenich, Creator of Aravrit

Meet Liron Lavi Turkenich, an entrepreneur and typeface designer from Haifa. Since earning her degrees in visual communications at Tel Aviv's Shenkar College and Typeface Design in the University of Reading in England, Liron has designed multilingual typefaces for companies, led workshops on design and entrepreneurship, and produced and curated an international conference for ATypI, the Association Typographique Internationale. But most fascinating of all (to us, at least) is Liron's creation of Aravrit, a hybrid writing system and portmanteau or combination of the Hebrew words *Aravit* (Arabic) and *Ivrit* (Hebrew). Here, we talk to Lavi Turkenich about what attracted her to visual design, how her creation inspired people around the world, and how Aravrit inspired a woman to better appreciate her grandmother.

Benji Lovitt: I must say, Liron, that beyond being beautiful to look at, Aravrit is actually readable by a Hebrew-speaker like me. How does it actually work, logistically?

Liron Lavi Turkenich: Thank you! In Aravrit, each letter is composed of Arabic on the upper half and Hebrew on the bottom half. As a result, it can be understood by speakers of either language.

How did you come up with the idea for Aravrit?

It actually wasn't a single moment. It was more the culmination of something that had been percolating in my mind. When I would take the train to Tel Aviv for class, I would often sit across from passengers and wonder what they were reading. In a purely creative and hypothetical way, I would think, "What if we could somehow read the same thing from two different perspectives?"

That was the designer-artist in you talking.

Exactly. Growing up, I actually had zero background in art and design; my whole family are engineers. But as a kid I always loved letters, starting with a toy letter set I had. So very early on, I knew that I wanted to do this in life, even before I knew anything about the professional world. In my typography class at university, my classmates would sometimes complain during the process of designing and copying letters. I, meanwhile, was in heaven.

And how did this lead to Aravrit?

Living in Haifa, I would see Hebrew and Arabic script everywhere but couldn't read a single word of Arabic. One day, while staring at a street sign, I realized that I had completely ignored the Arabic text my entire life, as if it weren't there. I felt like despite the coexistence of our two communities, we were separate – you might say parallel but not touching. I wondered if I could create a typeface that would force people, including myself, to read the same thing at the same time.

And what made you think that was even possible?

In the nineteenth century, French ophthalmologist Louis Émile Javal discovered that, for the Latin alphabet, our brains only need to see the top half of words to read them. It happens to be the same for Arabic, but not for Hebrew – in Hebrew, we need only the *bottom* half. Once I figured that out, I knew I was on to something.

So how did you begin and what were the challenges?

Before I even began the design, the toughest part was overcoming doubt. People around me didn't think it was possible. When I submitted the idea as my graduate project, the faculty didn't want to approve it at first for the same reasons. Luckily, my friends and family believed in me.

And the process itself?

It was very tedious! With twenty-two letters in the Hebrew alphabet and twenty-nine in Arabic, I had to create 638 new letters, one for every possible combination. Not only did each one need to be easily read in each language, I needed to be loyal to and retain the essence of each: Arabic with its curvy, flowing letters and Hebrew with its sharp ones.

With zero knowledge of Arabic, I was starting from scratch, which was very humbling. I would solicit feedback from anyone around me, asking them to try reading the words – it was a good thing I had so much time commuting on the train. Ironically, or maybe not, the first word I created was *language – safah* in Hebrew, *lura* in Arabic. I'll never forget it. When I realized that someone could read it in each, I celebrated. I still celebrate today.

What kind of positive feedback did you receive?

So many different kinds! After my TEDx talk, an audience member confessed to me that he had a creative idea but didn't think it would be accepted because he was "just a programmer." Once he saw that "even a typeface designer" can effect change in the world, he too had hope. When I lead workshops with young school kids, I see them trying to connect letters themselves. What's important isn't the typography, it's that they're learning to be creative and not let boundaries limit them. In one workshop, an Israeli woman told me about her grandmother who had made aliyah from Iraq long ago. When the grandmother would speak Arabic, the woman would respond, "Speak Hebrew, I don't want to hear Arabic!" Only when the granddaughter discovered Aravrit did her family history suddenly click for her. She appreciated the importance of her grandmother's roots and how they connect us to our present identity.

Beyond these personal interactions, where did you see evidence of growth?

I began receiving Aravrit requests for logos, T-shirts, workshops for students. Murals and graffiti with Aravrit suddenly appeared in Jerusalem, Tel Aviv, and Haifa. It even made an appearance in President Reuven Rivlin's holiday video for Ramadan.

What about outside of Israel?

I've already done a few design projects in the UAE – greeting cards, the cover of a Jewish/Emirati cookbook – but the biggest was participating in Expo 2020, the world fair held in Dubai. Before the agreement was even announced, I was asked to design something for Israel's exhibit: a thirteen-meter-long Aravrit statue saying *"El hamachar"* (to the tomorrow). At the time, I didn't know if I'd ever actually get to see it live with my own eyes. After the announcement, I visited several times, which was fantastic. People from around the world would ask me, "It's not Arabic, it's not Hebrew, what is it?" That's exactly what it should do – spark discussion and curiosity. But more than that, Aravrit makes people feel *seen*, that there is value in their identity. And once that happens, they think, "What other connections are possible?"

What are your dreams for Aravrit going forward? Do you hope to get involved at a policy level?

I hope Aravrit continues to bring people closer together by helping them see things they hadn't noticed before. And like the programmer, I hope others bring creativity and new points of connection to their own lives. All of us are joined together and intertwined in this world, whether we like it or not. For now, I'm intentionally keeping it very organic, to make change at a grassroots level through my workshops and art. I love meeting people in countries around the world, inspiring them to create their own ideas. We don't need to wait for our governments to change the world.

So no street signs in Aravrit?

It sounds wonderful, but that's not something I'm pushing for. I'd hate someone to be confused on the highway and miss their exit!

CHAPTER 4

Government, Social Policy, and Education

THE MOST STABLE POSITION IN ISRAEL'S NATIONAL LEADER-ship is its president, the country's head of state, who serves in a mostly apolitical and ceremonial manner – similar to the king or queen of England, if teatime were replaced by convoluted coalition math. While tea is a pleasant leisure activity, coalition math is a thankless and challenging task by which the president chooses the eventual prime minister.

How thankless and challenging, you ask?

In 1952, Israeli ambassador to the United States Abba Eban asked Albert Einstein if he was interested in the position. Einstein declined, claiming he was not qualified. When the guy who created the Theory of Relativity doesn't think he can handle the job, that tells you something.

Israeli politics, and specifically the Knesset (Parliament), is not for the faint of heart. It's filled with politicians who communicate through screaming, insults, and the frequent banging of fists on furniture. Why so angry? Could it be the lingering stress from one of the many wars the country has suffered? The difficult economy? Or maybe it's the entire political system that's to blame. You'd feel pressure too if your government were shakier than a game of Jenga. Indeed, when we first wrote this section, we planned to explain why a country would need to hold four elections in two years...until it turned into five in three and a half.

Buckle up for chapter 4, in which we attempt to explain the bumpy roads of Israeli government. We'll also look at some of the pressing (and non-pressing) issues of Israeli social policy, as well as at the education system (don't worry, we've included time for recess).

Speaking of politics: In November 1947, Winston Churchill said that "democracy is the worst form of government, except for all the others that have been tried." Little did he know what was just around the corner...

Politics: 120 ÷ 2 + 1 = Chaos

In 2019, Israel held what some called "the most important election in the country's history." How important? So important that the country held four more over the next three and a half years.

Indeed, between April 2019 and November 2022, Israelis went to the polls five times. This sounds absurd, until you understand how the Israeli government works – or, as many Israelis would say, that it doesn't.

"The trouble begins with the fact that when we Israelis step into the voting booth, we don't vote for a person, but for a party," says Dr. Gideon Rahat, chair of the political science department at Hebrew University and senior fellow at the Israel Democracy Institute.

Yes, come election time you'll see posters and TV commercials for Bibi or Bougie or any other nicknamed candidate for prime minister. But when you cast your ballot, you don't actually vote for him or her; instead, you vote for your favorite candidate's party, with the tacit understanding that if that party is able to form a government and assume power, they'll choose whoever was on the poster to be prime minister.

Fair enough. But consider that in any given election, there are upwards of thirty political parties to choose from. "To qualify for a seat in the Knesset (Israel's parliament), a political party needs to receive at least 3.25 percent of the overall national vote," Rahat explains – what's known in Hebrew as the *achuz hachasimah* (electoral threshold). Parties that cross the threshold are allotted seats in the Knesset in proportion to the percentage of votes they received in the election – but only after all the other votes (for parties that didn't reach 3.25 percent) are thrown out.

As a result, many votes end up "not counting," because any party whose vote total doesn't cross the 3.25 percent threshold is simply tossed, similar to how voting for a third-party candidate in the US is considered throwing away one's vote. Picture a US presidential election with one Republican, one Democrat, and twenty Ralph Naders.

Still with us?

Good! So now we'll talk about the consequence of this system: for Israelis, voting is sometimes less about core beliefs and more about game theory.

For example: In your heart, you might agree with the progressive policies of left-wing Meretz. But if you don't think they'll get enough votes to cross the magic threshold, it'd be a waste to vote for them. Better to vote for your second choice, Labor.

But wait! Labor is pretty much *guaranteed* to cross the 3.25 percent threshold, so maybe it makes still *more* sense to vote for Blue and White, since they have the best chance to unseat Likud, which they can only do with the help of a coalition that includes Labor, which means that Labor's voice *will* be heard…

How do you say "undecided" in Hebrew? (There's actually no single word for it.)

But the biggest consequence of this system is that with so many small, single-issue parties in the running, no major party is able to attain a clean sixty-one-seat majority on its own in the 120-seat Knesset. "The closest we came was in 1969, with Golda Meir's 'Alignment' joint list," Rahat says. That only happened because Meir brought several parties together *before* the election (an alliance that would later be renamed Labor), winning fifty-six of the required sixty-one. To show you how mangled Israeli politics is, note that Meir's coalition happened right after the Six-Day War, which just goes to show that even when you save the country from annihilation at the hands of eight hostile countries, and you do it in under a week, it's still virtually impossible to form a coalition.

Anyway…

You can guess what happens next: with no single party ever able to win a majority of Knesset seats, the fate of the country suddenly falls into the hands of a few small parties who demand concessions for their loyalty from the would-be prime minister.

Let the political blackmail – er, games – begin! "Yes, we'll join your coalition," leaders of these fringe parties say. "But only if you undo the Western Wall reform that would allow egalitarian prayer, give us no fewer than three prestigious ministries, and make the Tefilat Haderech (the Traveler's Prayer) an official part of the public elementary school street-safety unit."

Done!

Usually, it's small religious parties like Shas or United Torah Judaism (UTJ) who are the kingpins. That's why the Orthodox-controlled Rabbinate has such powerful influence over marriage and life-cycle events, prevents El Al from flying on Shabbat, and allows army exemptions for eighteen-year-old ultra-Orthodox Jews who serve the country by studying in yeshivas.

It's also why Cabinet positions are often filled by people who have no relevant expertise. One egregious example was in 2020, during the COVID-19 pandemic, when the minister of health was Rabbi Yaakov Litzman of the religious UTJ – a man who not only knew nothing about medicine, but was against vaccines, masks, and social distancing. His belief was that God would save people from the virus. (Interestingly, Litzman eventually got COVID and soon after resigned, thereby suggesting that God apparently preferred to sit in the opposition.)

Israel's system also encourages Cabinet members to jump from one post to another, simply as a way for the prime minister (who assigns Cabinet portfolios) to keep certain influential politicians in the government. After the 2015 election, Prime Minister Benjamin Netanyahu named fellow Likud member Miri Regev minister of culture and sport, and in 2020 made her the minister of transportation as well as agreeing to promote her to foreign minister later. With just seven more holes in her punch card, Regev would be the lucky recipient of a free ministry.

This all points to a certain level of dysfunction. So how might Israel break out of this cycle?

Rahat offers a suggestion: the prime minister should be chosen from the party that gets the most votes, instead of from whichever parties can form a coalition. "The benefit of this system is that it would make it harder for small parties to extort," he explains. "The PM could rule with a minority government and would still have to build and rebuild majorities for its policies."

And if people aren't happy with the outcome?

"Not a problem," Rahat says. "We'll just vote again."

Who'd I Just Vote for, Anyway?

In one of the most recognizable moments of Jewish culture, Tevye the Milkman sang about tradition while standing next to a horse.

Tradition is and always has been an essential ingredient to Jewish life, and Israel's government is no different. Though Israel is considered both a liberal and parliamentary democracy, complete with plenty of official rules and procedures, at the heart of Israeli government is a unique system of traditions that make it unlike any other in the world.

Here are twelve customs, practices, and interesting facts that make Israeli government one of a kind...

1. **What's old is new: the Knesset.** The name of Israel's parliament (the Knesset) and the number of members (120) are both derived from the ancient Knesset Hagedolah (Great Assembly), a group of 120 scribes, sages, and prophets who governed the Jewish people

beginning after the age of the prophets and continuing until the birth of rabbinic Judaism (from about the sixth to fourth centuries BCE). Popular representatives include Ezra, Zechariah, and Judah Hanasi, best known for compiling the Mishnah – a far greater accomplishment than anything done by Knesset members today.

2. **Prime minister for life.** Israeli think tanks and even the Knesset itself have discussed reform, but for the time being Israeli politicians can serve as prime minister for as long as people are willing to vote for them. Not that Israelis actually *vote* for prime minister…

3. **So, uh, who exactly am I voting for?** On Election Day, Israeli voters choose a party, not an individual candidate. Aside from each party's leader, who will then go on to become prime minister if the party wins enough seats to form a government, the average voter may not even know any of the other names on the party list. The only exceptions were the elections of 1996 and 1999, when voters chose a party *and* a prime minister, and the prime ministerial election of 2001, which took place with no Knesset election.

4. **Democratically elected… sort of.** If voters choose a party, who's choosing the people *in* the party? While some parties hold primary elections to determine the party list (the ranked order of politicians who would serve in the Knesset), most don't; their lists are chosen by their party leadership.

5. **Taxation without representation.** The only voting district in Israel is Israel. So on the one hand, you never have to worry about gerrymandering or disproportionate representation based on geography. But on the other hand, no member of Knesset is beholden to any one group of constituents. Moreover, because representatives aren't chosen via direct elections, it's not possible to lobby Knesset members or hold them responsible for their performance. Between this and point number four above, you see why the Israeli electorate is unable to mimic American voters who aspire to "throw the bums out" of Congress (Israeli party leaders are too busy throwing them in).

6. **They love to argue but will never debate.** When elections are between competing parties, not persons, it follows that there are no debates between leaders in which they argue the issues. Though televised debates have taken place in the past, the last one to include the prime minister was in 1996, the same election that featured direct voting. The winners? The politicians, who don't need to clearly articulate a platform of beliefs and policy goals. The losers? The electorate.

7. **"I'm running for Knesset, and I don't even have the chance to approve this message."** Except for specific allotted times, the Knesset Election Law forbids parties from advertising on TV and radio in the sixty days before elections. But don't fret – politicians will still find you, most likely through your cell phone, which will be bombarded with unsolicited text and WhatsApp messages up to and through Election Day.

8. **Election Day BBQ!** Israel is among the countries of the world where Election Day is a national holiday. So no, you won't get to prance into the office sporting an "I Voted!" sticker, but neither will you have to race to the local polling station at seven a.m. to beat the crowd – or worse, decide (gulp) not to vote because you couldn't find the time. Between time off from work and school and the chance to take a day trip with family and friends, Election Day actually feels a lot like Independence Day, except instead of coming before, the mourning comes after.

9. **Voting with your feet (or wheels).** In an effort to ease and encourage voting, the Central Elections Committee makes intercity public transportation service free for those whose voting district is in another city as determined by the address listed on their *te'udat zehut* (national ID). Those who have to travel to the southern resort and sales-tax-free city of Eilat must register and provide proof of eligibility in advance – which, yes, is a pain, but it saves the country the cost of transporting millions of Israelis who are willing to lie in exchange for a free beach vacation.

10. **Thanks, and enjoy your ministry!** As mentioned in the previous chapter, to keep coalition allies loyal, prime ministers sometimes dole out Cabinet positions as prizes – and if there aren't enough Cabinet spots to go around, they simply make up new ones. Case in point: in 2019, Benjamin Netanyahu created the "Ministry of Cyber and National Digital Matters" – a ministry so valuable and robust that it was eliminated two years later.

11. **"We, the people..."** Israel has no written constitution. Instead, its Chukei Hayesod (Basic Laws) serve a similar purpose and help guide the courts in their decision making.

12. **Pushing the envelope.** The 2000 US presidential election was thrown into chaos when thousands of voters in the heavily Jewish State of Florida either didn't successfully punch through their cards (the "hanging chad" controversy) or mistakenly voted for the wrong candidate (the "butterfly ballot" controversy). Fortunately, voters in the heavily Jewish State of Israel need not worry about a similar catastrophe, thanks to a ballot system that may have been designed by a first-grader running for student council. Here's how it works:

    Step 1: Walk into booth.

    Step 2: Find slip of paper with your party of choice's name on it.

    Step 3: Pick up slip of paper with your hand.

    Step 4: Insert slip of paper into envelope.

    Step 5: Drop envelope into box.

    Step 6: Feel good about exercising your democratic rights on your day off from your programming job in high-level cybersecurity.

Yes, the paper-in-an-envelope system seems very rudimentary. On the other hand, this system is very unlikely to be hacked by Russian or other foreign powers.

Prime Minister Cheat Sheet

Quick: How many Israeli prime ministers have won the Nobel Prize? Which prime minister served for less than a month? More than half of Israel's PMs at some point Hebraized their last names, assuming office with a name other than the one with which they were born – true or false?

Here's your handy cheat sheet full of facts and stories about Israel's PMs, including some you've probably never heard before...

DAVID BEN-GURION (né Guren), born 1886 in modern Poland

Term: 1948–1954, 1955–1963, Mapai Party

Reality check. Israel's longest-serving prime minister until Benjamin Netanyahu, Ben-Gurion famously said, "In Israel, in order to be a realist, you must believe in miracles."

When not in office, you might find him... Doing one of his famous headstands on the beach.

What's in a name? Both the international airport and university in Be'ersheva are named for him, and just about every town in Israel has at least one Ben-Gurion Street/Boulevard/Park. So if you tell a cabbie to get to Ben-Gurion on the double, be specific.

MOSHE SHARETT (né Chertok), born 1894 in modern Ukraine

Term: 1954–1955, Mapai Party

Hello, I must be going... After Ben-Gurion "retired" from politics, Sharett took over as PM. Upon Ben-Gurion's return the following year, Sharett stepped down, having served less than twenty-four months.

Special talents. Fluency in Arabic and Turkish allowed him to serve as an interpreter in the Ottoman Army.

They don't make 'em like they used to. From 1988 to 2017, Sharett appeared on the twenty-shekel bill.

LEVI ESHKOL (né Shkolnik), born 1895 in modern Ukraine

Term: 1963–1969, Mapai/Alignment Party (later renamed Labor Party)

Make yourself at home. Eshkol was a founding member of Kibbutz Degania Bet, one of the oldest kibbutzim and destination of many elementary school field trips.

The writing on the wall. While living in Berlin in the 1930s, Eshkol negotiated the Haavara Agreement with Nazi officials, allowing Jews fleeing persecution to transfer assets to British Mandatory Palestine.

See you at the top... Thanks to the archaeological excavations that began during Eshkol's term, you can enjoy the sunrise on Masada.

YIGAL ALON (né Peikowitz), born 1918 in pre–British Mandate Palestine

Term: 1969, Labor Party

Living out of his suitcase. Alon served as the interim PM for nineteen days after Levi Eshkol died of a heart attack in office.

That's my name, don't wear it out. Named for its architect, the "Alon Plan" proposed that Israel would cede much of the West Bank to Jordan following the Six-Day War while retaining strategic portions along the Jordan River for security.

GOLDA MEIR (née Mabovitch), born 1898 in modern Ukraine

Term: 1969–1974, Labor/Alignment Party

Bragging rights. Long before she became Israel's first and only female PM, Meir received the very first Israeli passport when she departed for Moscow in 1948 to serve as ambassador.

Walking in her footsteps. Her clunky orthopedic shoes became known as *na'alei Golda* (Golda's shoes) and were part of the female soldier uniform for many years.

More than beer and bratwursts. Golda grew up in Milwaukee, Wisconsin, her family fled pogroms in Ukraine.

YITZHAK RABIN, born 1922 in Mandatory Palestine under British rule

Term: 1974–1977, 1992–1995, Alignment/Labor Party

Local boy makes good. The first elected PM to be born in the land of Israel, Rabin won the Nobel Peace Prize for the Oslo Accords and signed a peace deal with Jordan during his second term.

And he didn't even have Instagram. As chief of staff, Rabin led ground forces into the Old City of Jerusalem during the Six-Day War – a moment captured in one of the country's most iconic photos.

In memoriam. After he was assassinated at a peace rally in Tel Aviv's Kikar Malchei Yisrael (Kings of Israel Square) in November 1995, the location was renamed Kikar Rabin.

MENACHEM BEGIN, born 1913 in modern Belarus

Term: 1977–1983, Likud Party

You say you want a revolution! Begin broke the left's twenty-nine-year stronghold on the government in what became known as the *mahapach* (revolution). The next year, he signed the Camp David Accords with Egyptian president Anwar Sadat, winning the Nobel Peace Prize.

Those were the days... Many older Israelis look back on Begin as the ultimate prime minister, reminiscing about him the way older Americans might about Kennedy or Reagan.

D'oh! In season 5, episode 9 of *The Simpsons*, Bart is prescribed a pair of square-framed black eyeglasses. His optometrist says, "Menachem Begin wore a pair just like them."

YITZHAK SHAMIR (né Yezernitsky), born 1915 in Belarus

Term: 1986–1992, Likud Party

Fight or flight. During the British Mandate period, Shamir joined and eventually led Lechi, the paramilitary organization whose main objective was to oust the British from Palestine.

Words of a prophet. According to Ehud Olmert, when Shamir's father, living under Nazi occupation, was informed that the extermination of the Jews was imminent, he replied, "I have a son in the Land of Israel, and he will exact my revenge on them."

Never forget. While working for the Mossad in the 1960s, Shamir directed Operation Damocles, resulting in the assassination of former Nazi scientists in Egypt.

SHIMON PERES (né Perski), born 1923 in Poland (today's Belarus)

Term: 1984–1986, 1995–1996, Alignment/Labor Party

Shhh... Peres established the Dimona nuclear reactor as architect of Israel's secret nuclear weapons program in the 1960s, partly with the help of Arnon Milchan, Hollywood producer of *JFK* and *Fight Club*.

They say Facebook is for old people... Peres was lauded for his social media presence and ability to communicate with the public, including his comedic YouTube video "Former Israeli President Shimon Peres Goes Job Hunting," and to promote causes he believed in, such as coexistence with Palestinians.

Sharing is caring. For his work on the Oslo Accords, Peres shared the Nobel Peace Prize with Rabin during his second term.

BINYAMIN "BIBI" Netanyahu, born 1949 in Israel

Term: 1996–1999, 2009–2021, 2022–the time of this writing, Likud Party

Move over, David. Netanyahu is Israel's longest-serving prime minister and the first to be born in the State of Israel.

Me! Choose me! In 1996, Netanyahu defeated Shimon Peres in the first direct election in which electors voted for PM and Knesset separately.

Say my name. While earning a bachelor's degree in architecture and a master's from MIT, Netanyahu went by the easier-to-pronounce name Ben Nitay.

EHUD BARAK (né Brog), born 1942 in Mandatory Palestine

Term: 1999–2001, One Israel/Labor Party

Man of many medals. Barak is Israel's most decorated soldier, having served in the IDF for thirty-five years and as chief of staff.

Glass half something. Speaking to the foreign press, Barak opined, "They say in the Middle East a pessimist is simply an optimist with experience."

Brains *and* brawn. A trained classical pianist, Barak also earned degrees in physics, mathematics, and economic engineering systems.

ARIEL SHARON (né Sheinermann), born 1928 in Mandatory Palestine

Term: 2001–2006, Likud/Kadima Party

No more Mr. Nice Guy. Sharon was nicknamed "the Bulldozer" for his tough and aggressive nature.

Not a moment too soon. Sharon's military leadership turned the tide of the Yom Kippur War as he led the 143rd Armored Division across the Suez Canal and encircled the Egyptian Third Army.

Say cheese! One of the most indelible images of Sharon is a photograph from his ranch with a sheep draped around his neck.

EHUD OLMERT, born 1945 in Mandatory Palestine

Term: 2006–2009, Kadima Party

Thrown into the deep end. After Sharon suffered a stroke in office, Olmert became interim PM before calling for and winning elections.

He said it. "Leadership is tested not by one's capacity to survive politically but by the ability to make tough decisions in trying times."

I believe the children are our future. At age twenty-eight, Olmert became the youngest person until then ever elected to Knesset.

NAFTALI BENNETT, born 1972 in Israel

Term: 2021–2022, Yamina/New Right Party

You can't please all of the people all of the time. Bennett led a historically diverse "change government," the first to include an Arab party, bringing an end to "King" Bibi's twelve-year reign. One year into his term, he ceded the premiership to his rotation partner Yair Lapid when his government collapsed.

Show me the money. After his service in the IDF, Bennett moved to New York and cofounded a software company that eventually sold for $145 million.

Quiet, please! While the permanent PM residence underwent renovations for the duration of his term, Bennett ruled from his personal home (just a few blocks away from that of Joel, co-author of this book). Anti-government protestors often marched through the neighborhood banging drums, shouting into bullhorns, and playing kazoos.

YAIR LAPID, born 1963 in Israel

Term: 2022, Yesh Atid Party

Please, you first, I insist. Lapid engineered the change government that ended Netanyahu's reign, agreeing to let Bennett serve first in rotation despite receiving fewer votes than Lapid.

From the press room to the podium. Lapid spent decades as a reporter for the IDF, newspapers, and eventually television as the anchor of popular newsmagazine *Ulpan Shishi* (Friday Studio).

Lapid-er, Paul and Mary. A lover of the arts, Lapid wrote chart-topping pop songs for some of Israel's biggest singers.

As of this writing, Netanyahu is prime minister of Israel once again. Only time will tell how his legacy will be remembered as not only the longest serving PM, but also the one whose judicial reform proposal led to mass weekly protests and talk of civil war, and whose government was in power during the October 7th massacre.

As you now know, the answers to the questions posed at this section's beginning are three; Alon; and true.

But be warned – there's a chance that by the time you read this book, the list will be out of date. Did you *read* the last section, about how Israeli politics works (or doesn't)?!

Gay Rights: Pride in the Living Room

In June 2019, award-winning actor Neil Patrick Harris visited Israel as the Tel Aviv Gay Pride Parade's international ambassador. A quarter of a million locals and tourists partied, danced, and showed their support as colorful floats with revelers in various states of undress proceeded from Ben-Zion Boulevard to Charles Clore Park.

Between its hospitable beaches, ubiquitous rainbow flags, and a population thought to be almost a quarter LGBTQ, it's no wonder some have called Tel Aviv the number-one gay travel destination in the world. Still, Tel Aviv neither is perfect nor represents the entire country. While the Jewish state has made great advances on the LGBTQ front, rabbinical control over life-cycle events (covered in chapter 2) creates complications for all its citizens. After the parties die down and the tourists head home, what is daily LGBTQ life like in Israel?

"A great parade doesn't mean the battle is over," says Jay Rosen, an American *oleh* (immigrant) and long-time volunteer with Israel Gay Youth, Israel's biggest LGBTQ organization. "Even Tel Aviv isn't immune to homophobic attacks." In one 2019 incident, a sixteen-year-old boy was stabbed outside an LGBTQ youth shelter, and 2021 recorded the highest number of anti-LGBTQ discriminatory acts in Israel's history, with a report of a hate crime received every three hours. "Activists and community members have unending work to do."

Anti-LGBTQ discrimination in the Middle East goes back a long way. In the 1500s, England passed the Buggery Act criminalizing same-sex relations and spent the next five hundred years spreading homophobia throughout its colonies, including Mandatory Palestine, where local laws decreed these relations punishable by death.

After Israeli independence in 1948, cultural awareness slowly increased through books, magazine articles, and the country's first gay bars. In 1963, attorney general Haim Cohn declared that sodomy laws would not be enforced (though they remained on the books until the late eighties).

The 1970s saw the creation of what is now Israel's oldest LGBTQ organization, the Aguda, central in the advancement of the gay rights movement, especially in the early nineties when they helped pass anti-discrimination legislation in the workplace. Soon after came the first Tel Aviv Pride Parade and the Knesset's abolition of restrictions on gays in the military. Ironically, this ruling coincided with the US Congress passing "Don't Ask, Don't Tell," which prohibited openly gay soldiers from serving.

Rosen explains, "Tiny Israel lacks America's issue of states' rights. It's easier to effect change when it only requires a single vote. Also, not to sound cynical, but in such a small place when we were fighting for our lives, we can only discriminate so much. We need as many productive citizens as possible!"

That includes a famous trailblazer who helped boost her community's visibility: Dana International, the winner of the 1998 Eurovision competition. If winning alone weren't enough, Dana being transgender did wonders to advance the LGBTQ cause.

Different parts of the country have progressed at different rates, though.

Haifa has a thriving LGBTQ community; Arabs and Jews have lived there in relative harmony for years, so adding another identity to a city with coexistence in its DNA is easy. Rishon L'Tzion, Israel's fourth-largest city, also ranks high on the Aguda's annual index.

The communities that score the lowest are typically the most insular, such as Bat Yam, a suburb just a stone's throw from Tel Aviv. What makes Bat Yam so different from its "gaybor" (Rosen's word) to the north? In part, the fact that approximately 30 percent of its population immigrated from Communist Russia, not exactly a bastion of free thinking and open ideas.

Unlike Tel Aviv, the attitude of Jerusalem's LGBTQ community is quality over quantity. The Jerusalem Open House has been a neighborhood pillar since 1997, and its Pride parade, established in 2002, marks the official beginning of Israel's Pride Month. Tel Aviv's may have more dancing, drinking, and inflatable penises, but Jerusalem's modest event symbolizes that the movement is about more than just fun.

In 2007, Avi Rose and his husband became the first same-sex couple in Jerusalem whose marriage earned legal recognition. "In Tel Aviv, the parade is about freedom and sexiness. It's international and not even seen as an *Israeli* event," Avi explains. "Jerusalem's is a serious political statement. It's about the complexities of being gay in Israel and owning your queerness in a Jewish space, and even, to a lesser extent, Muslim and Christian spaces."

For some, part of that complexity is balancing gender, sexual, and religious identities. Observant Jews who put their religious identity first may struggle in Tel Aviv, where the LGBTQ events and parties often happen on Shabbat. But the process of blending multiple identities is what makes Jerusalem special.

Social entrepreneur and LGBTQ community leader Nadav Schwartz echoes Rose's sentiments about integrating into the larger Israel. "Tel Aviv's message is 'We want rights even though we're different.' Jerusalem's is more 'We want rights because we're like you.' Within the 'Tel

Aviv bubble,' it's easy to isolate yourselves in a way that we can't in Jerusalem. We want to remain part of the communities and shuls we grew up in."

Successes aside, obstacles still remain. "Israel is a very family-oriented society," says Rosen. "It's not just the ultra-Orthodox in a rush to settle down." Even in the "forever young" environment of Tel Aviv, he explains, many people in the LGBTQ community are looking to partner up and have children. "This heteronormative desire for a family has trickled down to us. It speaks to how incorporated we are in larger society but also presents challenges."

Indeed, the community's latest struggles focus around unions, marriage, and parenting, making 2021's ruling to allow same-sex partners and single men to become parents through surrogacy within Israel a major achievement (though, as Rosen notes, it will take time for the systems to catch up and function smoothly). And whereas Tel Aviv's municipality grants the same benefits to same-sex couples as to mixed, this does not extend nationally.

"Tel Aviv is too expensive and claustrophobic to support every LGBTQ citizen," says Rosen. "We need anti-discriminatory laws and comprehensive recognition of same-sex marriage nationwide so we can live openly and be accepted anywhere."

"It's a complicated fight," says Schwartz. "Of course I want equal marriage. But because of the Rabbanut, it's not just an LGBTQ issue."

To make these ideals a reality, the movement has recently taken a page from another playbook. Since 2011, millions of Israelis gather in friends' living rooms on the eve of Holocaust Remembrance Day for Zikaron B'Salon (Remembrance in the Living Room) to hear survivors tell their stories. Recognizing the value in this kind of open, honest conversation, the new Israeli phenomenon called Ga'avah B'Salon (Pride in the Living Room) lets speakers share their personal struggles to break down barriers and create a more inclusive society.

And as Israelis say, *"l'at, l'at"* (slowly, slowly), things are changing. In a December 2023 landmark decision, the High Court ruled that same-sex couples may now adopt children.

The movement's struggle for equality is not without some irony. "In my kids' school, my husband and I are somewhat of an anomaly," says Avi Rose. Not because they are gay, but, as he says with a smile, "We're one of the few couples who are still married."

Hot- (Or Not) Button Issues

Abortion. The death penalty. Gun control.

Bring up one of these topics at a dinner party in the US and you're guaranteed a lively (if not volatile) conversation, even if all the attendees agree.

Where does Israel stand on these hot-button issues? And what do Israelis have to say about them?

To answer the second question first, not much. We'll explain why in a bit. But first, your policy brief:

Abortion

In Israel, abortion is legal and, when all is said and done, accessible to just about any woman who wants one. The only hitch is the "all is said and done" part: To get an abortion, women need to plead their case before the local Va'ad Ishur Hafsakat Herayon – the Termination of Pregnancy Committee. The three-member committee consists of a gynecologist, a second medical professional, and a social worker, at least one of whom must be a woman.

Until June 2022, interviews were face to face. But days after the US Supreme Court reversed Roe v. Wade, Israel relaxed its abortion rules to make the process less burdensome. As of this writing, abortions can now be performed at local health clinics, not just in hospitals, and in-person interviews have been replaced by a digital application. "We've been advocating for these changes for years," says Sharon Orshalimi, a reproductive health and rights scholar and activist. She notes that the application still constitutes an invasion of privacy. "Women have to answer questions about birth control and their sexual history. Which are totally irrelevant."

The rest of the abortion law remains mostly unchanged. Women under eighteen are approved automatically, with no parental consent required; so are women over forty and anyone whose pregnancy is the result of rape, incest, or what's known as an "illicit relationship."

The one group whose petitions are sometimes rejected are married women between eighteen and forty. Women in this category need to convince the committee of a "valid" reason. "To avoid problems, these women will sometimes tell the panel that they committed adultery or are 'mentally unwell' and can't handle the demands of the child," says Shoshanna Keats Jaskoll, cofounder of Chochmat Nashim (The Wisdom of Women), an Israeli nonprofit that advocates on behalf of women.

Still, with the over 98 percent of requests approved, many feel that the committee is unnecessary: If the interview is a formality, why subject women to an intrusive grilling?

Gun Control

At first glance, you'd be excused for assuming that Israel lacks gun control. First-time tourists are often astonished by all the soldiers, both in uniform and in street clothes, walking around with assault rifles. It's almost impossible to enter any bus station and not see a young man or woman with an M-16 slung casually over their shoulder, often equipped with a sniper scope or grenade launcher.

But as Hebrew University professor of law and criminal justice Dr. David Weisburd explains, Israel has very strict gun control laws. "Guns are highly regulated in Israel," he says. "Having a gun is expensive and requires a good deal of licensing." To own a gun, Israelis must prove they *need* one – either because of where they live (typically over the Green Line in the West Bank) or for their profession (such as security guards, tour guides who take groups into dangerous areas, or veterinarians who work with large animals and need protection). Even then, according to the *Times of Israel*, about 40 percent of gun applications are rejected; approved applicants can own only one gun at a time, and almost always a handgun or pistol – no AR-15s or other automatic weapons for the masses.[1] Ammunition is limited to fifty bullets per purchase.

Beyond just the law itself, Weisburd notes, the Israeli attitude toward guns is very different from that of many Americans. "Israelis have experience primarily through the army," he says. "Though young recruits 'enjoy' the idea of having a gun in the army at first, they soon come to see it as a responsibility. For most of the Israeli population, guns are not for leisure but a necessity for national protection."

Capital Punishment

In theory, Israel allows for the death penalty in cases of treason, genocide, or crimes against humanity. In practice, Israel has used the death penalty just once: In 1961, Nazi war criminal Adolf Eichmann, mastermind of the Final Solution, was tried and convicted in a Jerusalem court and subsequently put to death. Though technically still on the books, capital punishment is almost never sought by Israeli prosecutors, even in cases of terrorism. In a paper titled "Israel and Capital Punishment," the late Hebrew University law professor Dr. Edna Ullmann-Margalit explains that Israel's attitude toward the death penalty is rooted in Judaism:

> Despite various biblical injunctions in the spirit of "Whoso sheddeth man's blood, by man shall his blood be shed" (Genesis 9:6), normative rabbinic Judaism recoiled from the death penalty. Indeed, the rules of evidence adopted by the Sanhedrin were so strict as to effectively prevent [it].[2]

Ullmann-Margalit also points out the modern, practical considerations for why Israel doesn't resort to capital punishment, including a lack of evidence that it serves as a deterrent and fears that executing terrorists might lead to revenge killings and kidnappings.

---

1. Haviv Rettig Gur, "Comparing America to Israel on gun laws is dishonest – and revealing," *Times of Israel*, Mar 1, 2018.
2. Edna Ullmann-Margalit, "Israel and Capital Punishment," *Sh'ma* (Oct 2002).

So why are these three hot-button issues, each so polarizing in the US, rarely talked about in Israel?

It's because, for the most part, Israelis don't feel the government is trampling on their constitutional rights. (It probably helps that Israel doesn't have a constitution.)

But there's another reason: tradition. As Weisburd explains, "Judaism has always been averse to the death penalty – the rabbis added so many stipulations before capital punishment could be carried out that it was extremely difficult to actually sentence someone to death. The Talmud notes that a Sanhedrin (Supreme Court in the Second Temple era) that puts to death one person in seven years is termed tyrannical, and Rabbi Eleazar Ben Azariah took it further, saying 'One person in seventy years.'"

There's no gun culture in Israel, either: Israelis don't hunt, and you won't find "Guns & Ammo" on the shelf of your local bookstore, in large part because *kedushat chayim* – the sanctity of life, for both humans and animals – has always been a cornerstone of Jewish thought. In general, Israelis have never considered guns as anything other than a way to protect themselves.

As for abortion: "Even fifteen hundred years ago," Weisburd says, "these same rabbis decided that the mother's health must take precedence over the unborn child's when her life is at risk."

These traditions, though rooted in ancient behavior and ideas, are nevertheless the foundations for how modern Israelis think about these issues.

Or, more accurately, don't.

Street Safety: "Pretend She's Your Grandma"

When was the last time someone gave you advice on how to cross the street?

If you're an Israeli who listens to the radio, the answer is, "An hour ago." At the top of the hour, around the clock, Israeli radio stations run public service ads (PSAs) reminding pedestrians to make eye contact with drivers before stepping into the street and imploring drivers to wear seatbelts, never text and drive, and to please, *please* relax and not throw a fit if someone cuts them off in traffic.

As we explored in earlier chapters, Israelis have evolved to never be *freiers* and to bend the rules when necessary. We also discussed that people carry tremendous stress due to economic conditions and past and present threats of conflict. When these stressors manifest via heavy equipment moving at high speeds on the road, the consequences can be fatal. Smartphones have made both pedestrians and drivers ever more distractible, and to make matters worse, police barely enforce the laws.

Because Israeli culture values risk taking and bold, out-of-the-box thinking, don't be surprised to see a driver react to an idle garbage truck in the road by passing him on the sidewalk or driving in reverse to the intersection behind.

Hence the PSAs: tiny, hourly reminders that the lives of all Israelis are in *our* hands.

Dr. Moshe Becker, director of transportation studies at Galilee College and one of the country's leading experts on road safety, views these PSAs as a welcome change, but not enough. "A smart person knows how to get out of a problem," Becker says. "The wise person knows how to avoid the problem in the first place."

Becker, who regularly advises the Knesset on transportation, wants Israel to shift from reaction to prevention. "Israelis are great at *chilutz v'hatzalah* [search and rescue]," he says. "What we need to do is stop these accidents before they happen."

One place to start is infrastructure. "We're building new roads at a rate of 1.5 percent a year," he says. "Meanwhile, the number of vehicles on the road is going up by 4.5 percent. That's too many cars on too few kilometers of road."

In addition to new and safer roads (he notes that those along the Dead Sea are particularly dangerous), Becker would like to see more underground transport and increased police enforcement. "At any given moment, we have two hundred police cars on the road," he says.

"That's in the entire country." For the most part, these officers give tickets, instead of patrolling the roads for errant driving and stopping wrecks from happening.

According to Becker, Israel experiences about five hundred thousand traffic accidents per year (more than thirteen hundred a day), eighty thousand of which lead to injury or death. "These accidents cost Israel more than 17 billion shekels a year," Becker says. "If we spent even 10 percent of that on improving roads, public transport, and media campaigns, we could significantly decrease the number of accidents. And deaths."

In the meantime, the Ministry of Transportation's PSA campaign continues to run on the hour and appears on billboards and even sidewalks around the country.

Here are some of the most popular recent campaigns.

"It's not gonna kill you." A popular Hebrew expression is *"Af echad lo yamut"* (Nobody's going to die). It basically means, "It's no big deal, it's not gonna kill you." One billboard campaign uses the phrase literally, declaring "Slow down, nobody's going to die." Unless they continue driving recklessly, of course.

"Look 'em in the eyes!" In these billboard and radio PSAs, pedestrians are advised to make eye contact with drivers before they step into the crosswalk. Posters feature a set of large, ominous eyes to drill home the message.

"Pretend it's your grandma." These radio ads ask drivers to imagine that pedestrians crossing the street are their grandparents, closest friends, and young siblings. "You'd stop if it were your grandmother!" the ads cry. "So why not stop for other pedestrians, too?"

"Stop, Look, Eyes, Go." These step-by-step instructions for how to cross the street are often painted next to the crosswalks near schools. They remind kids not to cross until the driver has acknowledged them.

"Put down the phone!" This campaign specifically targets girls and young women. Yes, it sounds sexist, but apparently the Ministry of Transportation believes too many girls and women are crossing the street while looking at their phones instead of at traffic.

Are these PSAs working?

A bit, Becker says. But he wants to see more.

"You know how when you watch basketball on TV, the commentators analyze exactly how the players move around the court by diagramming plays on the screen?" he says. "We should do that with car accidents. Every night on the news, show the 'Accident of the Day' with video and photos from the scene. Experts would explain how it could have been avoided.

"Finally," he says, "citizens of this country would realize that their lives are in their own hands."

Better our lives than our phones.

Field Trip to Yemen: School Daze

When Joel's son was in second grade, his substitute teacher threw a pencil at his face, just missing his eye.

Had this happened in Westchester, New York, where Joel used to live, there would have been a lawsuit, press conference, and letter from the head of school to all parents reassuring them that *violence toward the students will not be tolerated!*

But in Israel, Joel found *himself* on the defensive. "Your son was goofing off," the principal said.

"I realize that," Joel said. "But what about the teacher? He could have taken out my son's eye!"

"*Teraga* [relax]," said the principal. "I spoke to the teacher, and it's all just a big misunderstanding. He was aiming for your son's chest, not his eye. Like I said, your son has trouble sitting still."

Now, the vast majority of Israeli teachers are wonderful people who've dedicated themselves to the thankless task of educating youth and do *not* chuck writing utensils at pupils' pupils.

That said, the classroom is a microcosm of Israeli society itself: a place where rules are flexible, strong boundaries don't necessarily exist, and authority figures aren't respected the way they might be in other countries. Parents don't get too involved in their children's schoolwork, which lines up with the prevailing parenting philosophy in Israel – it's a fairly hands-off enterprise.

Yet somehow, Israel produces a disproportionate number of world-class scientists, writers, academics, and entrepreneurs. It boasts more Nobel laureates per capita than the US, France, or Germany. The only countries with more companies listed on the Nasdaq are China and the US.

How is this possible?

In part, it's because education is about more than just what you learn in school. Some of it is a matter of statistics: no matter how good or bad a country's school system is, some students are bound to go far and others to fail.

But according to Dan Senor and Saul Singer, authors of the best-selling book *Start-Up Nation*, there's another reason, one unique to Israel: they hypothesize that Israel's success in high tech is due in part to out-of-the-box thinking and chutzpah – traits that come easier when parents aren't hovering over your shoulder and rigid boundaries don't exist.

Here are a dozen must-knows about Israeli education:

That's right, *five*! Israel has five public school systems, all free: Jewish secular, Jewish religious, Arab Christian, Arab Muslim, and Arab Druze. In areas with more than one option, parents choose which school system to enroll their children in. Private schools exist too, but are rare (at least in the secular community). Most cost around 8,000 shekels (about $2,500) a year, although schools for diplomats' children could run into the high tens of thousands.

Call me Ruti. In accordance with Israelis' distaste for all things formal, students call teachers and even the school principal by their first names.

Holiday vacay! Kids get three major breaks during the school year: Sukkot (two weeks), Chanukah (eight to ten days), and Passover (two weeks), plus the two-month summer break known as *chofesh hagadol* (the big vacation).

Shall we schedule for Friday, 9:45–9:50? In most elementary schools, the day runs from 8 a.m. to 1:30 p.m., Sunday to Thursday, plus Friday until 11:30 a.m. or 12 p.m. Because high-tech and other white-collar workers typically have Fridays off, parents relish Friday mornings, as the one day of the week when they're off work and the kids are out of the house. Many parents reserve Friday mornings as "date time" – going to a café together, meeting friends for a bike ride, or reserving a block of time to fulfill the traditional mitzvah of "delighting in the Sabbath" by, well, you can figure it out…

Thou shalt study for thy test! Bible studies is a core curriculum subject in all schools through twelfth grade, though the version used depends on the school system. In Jewish schools, kids study the Hebrew Bible (Old Testament), whereas in Arab Christian, Muslim, and Druze schools, kids take Christian, Islamic, or Druze studies, respectively, which delve into the New Testament or Koran.

Strike that! Teacher strikes are common. Typical demands include increased pay, reduced class size, and, more recently, better legal protection against aggressive students.

No magic school bus. The *tiyul shenati* (annual class trip) and other field trips are the only times students ride together on a bus as a class. Though there are private van services, you won't see big yellow school buses like you do in America. Kids typically walk or carpool to school, or ride public buses on their own.

In the hot seat. Students typically attend parent-teacher conferences with their parents. Most of the conference is a conversation between teacher and child with the parents observing.

Outnumbered. Israeli classrooms are crowded. In many elementary schools, the student-to-teacher ratio is 30:1 or higher.

Uniforms, kind of. Schoolkids in public school don't typically wear uniforms per se, but they are required to wear a T-shirt, sweatshirt, or collared shirt featuring an iron-on insignia of the school. Come August, stores start advertising shirt-and-logo deals, typically five T-shirts for a hundred shekels. Kids dread the appearance of these signs: it means vacation is over.

House calls. Teacher salaries in Israel are low, beginning at $22,000 a year. Many teachers take advantage of the short school day by tutoring privately, usually for $55–60 per hour. They often need this second job just to make ends meet. As a rule, though, they are not allowed to tutor their own students.

Field trip to Yemen! The highlight of the school year is the *tiyul shenati*, a grade-wide excursion to a landmark, historical site, or nature spot somewhere in the country. Beginning in sixth grade, the *tiyul* is overnight – students, teachers, and lucky parent chaperones (it's the most coveted spot on the parent volunteer list – seriously!) sleep in Bedouin tents, youth hostels, or sleeping bags under the stars. At some point, students will serenade the bus driver with the traditional (albeit short) song:

> *Hanahag shelanu chev-re-man!*
> *Hu yikach otanu l'Tei-man!*

Translation:

> Our bus driver is a really great guy!
> He is going to take us all the way to Yemen!

It sounds better in Hebrew.

Of the many parents we interviewed, the most common assessment of their kids' schools was *"b'seder"* (it's okay); some went so far as to call it "babysitting," but those were extreme cases. Immigrants, particularly from North America, were shocked at the lack of homework and academic rigor. "In Advanced English, my kid's sixth-grade teacher would wheel in a TV and show them Disney movies," remarked one dismayed mom. "They should be reading *The Diary of Anne Frank*. Instead, they're watching *The Lion King*."

Still, parents don't seem overly concerned, either – certainly not as much as they admit they'd be in the States.

This is partially due to a lack of faith that anything could possibly change. But it also reflects the fact that Israelis simply don't engage in the high-stakes, cutthroat competition for college admissions that Americans do. Fourteen-year-olds don't spend their summer interning in a law firm so they can put it on their college application. They spend their summer...being kids. With the army only a few years away, the harsh realities of adulthood will come soon enough.

And, of course, this casual attitude is partially cultural: Why worry? As those students learned in English class, *hakuna matata* – which very closely resembles one of Israelis' favorite expressions: "*Yihyeh b'seder*."

It will be okay.

Start-Up Recess

Like so many Israeli institutions, the school system is strapped for cash. You don't need to read a financial report to see it; just stroll through a typical school. Buildings are sparse, and a typical Israeli classroom has at least thirty students for a single teacher.

The schools don't have playgrounds, either. A typical elementary school has a basketball court/soccer field, with a soccer goal situated under each basket, making for a chaotic scene with both sports taking place simultaneously. At Joel's son's elementary school in Ra'anana – considered one of the better towns in the country, socioeconomically – the "playground" consists of a gazebo built by the North American–based Jewish National Fund (JNF) and a vertical metal pole. (We assume for climbing.)

Given this, Israeli kids are left to invent their own games come recess time. Below are three of the most popular. These games are simple and don't require much equipment, but don't be fooled: they're fast-paced and action-packed.

Chayei Sarah (Life of Sarah). Named after the fifth Torah portion in the Bible (the one where, ironically, Sarah dies), Chayei Sarah is basically dodgeball, except with two twists. Instead of teams, it's every man for himself, and when someone catches a thrown ball, anyone who'd previously been knocked out by the thrower is allowed back in. This second twist keeps previously eliminated players actively involved, rooting for players who just minutes ago were enemies but upon whom they now depend to rejoin game play.

To play Chayei Sarah, all you need is a ball (the spongier the better) and a bit of empty land. A typical game lasts ten minutes, more than enough time for a class of fifth- or sixth-graders to play twice during a break. Injuries are not uncommon, particularly when the ball is old, and small pieces of sponge fly off into the eye.

What makes for a great Chayei Sarah player? "Being able to throw really hard and hit people in the face," profoundly expressed the sixth-grade athlete we consulted for this piece. "And also that you're a good catcher when people throw balls at your head."

Shalosh Maklot (Three Sticks). What do you get when you cross the Olympic triple jump with a class of high-energy fifth-graders with nothing to do? The answer: Shalosh Maklot, a long-distance jumping game that requires speed, explosiveness, and the ability to maintain one's balance when stepping on rounded objects.

To start, students lay three sticks, tree branches, or even yarmulke clips about a meter apart on the ground, parallel to one another. The first kid in line – let's call her Jumper One – takes a running start and then leaps over the first stick, then the second, then the third, all in one smooth, gazelle-like series of leaps and one-footed landings. Wherever she lands after her final leap becomes the new location for Stick Three; Stick Two is then placed equidistant between the first and third. Subsequent players must then leap the three sticks, in their new positions, the same way Jumper One did; those who can't are out. After everyone's taken a turn, Jumper One goes again, and wherever she lands after her third leap is the new location of Stick Three, and so on. The last player to successfully complete the course wins.

Though not a contact sport like Chayei Sarah, Shalosh Maklot produces its fair share of injuries, typically in the form of twisted ankles and scraped knees from botched landings. Our sixth-grader reports that his friend sprained an ankle after she landed on Stick Three, which just happened to be a cylindrical rod borrowed from the school sukkah.

"To win at Shalosh Maklot, you can't just be a good jumper," he says. "You have to know how to save your energy for when you jump over the last stick, and be good at not falling down if you step on one of the sticks."

One great feature of Shalosh Maklot is that it can be played indoors, making it ideal for Israel's rainy season.

Al Haserve (On the Serve). Think ping-pong without paddles, and instead of a ping-pong ball, a basketball, volleyball, or any other large ball that would bounce on a table. Ideally, Al Haserve is played on a ping-pong table (which many schools have outdoors, typically with a solid plastic net but no other table tennis equipment); an ordinary cafeteria table or even six desks arranged into a rectangle will suffice in a pinch, however, with string, a pencil, or the crack between desks serving as the net.

To start, Player One throws the ball over the net, and as it bounces on the table, both players say *"Al"* (Hebrew for "on"). Player Two returns the ball over the net with a one-handed combination slap/push, and when it bounces, both players say *"ha"* (the). Player One returns the ball, they call *"Serve"* upon the bounce, and it continues from there like ping-pong, with points earned when the opponent fails to successfully return over the net and on the table. First player to three points wins; the typical game lasts around a minute and a half.

The secret to premier Al Haserve play? "Really good aim," according to our sixth-graders. "When you can hit the edges and corners of the table, so the other guy doesn't have a chance to hit it back, that's good."

The most common injuries occur as a result of an unexpected ricochet directly into the face.

Gogo'im (Apricot Pits). This resembles a carnival game you may have played as a kid, but instead of tossing a bean bag through a hole in a box, kids throw dried-out apricot pits (preferably ones not from someone else's mouth). First, you cut different-sized holes in boxes and assign each hole a point value based on level of difficulty. The smaller the hole, the larger the point value. Each player takes a turn and tries to earn as many points as possible. The first to earn the agreed-upon number of points wins. While not officially verified, it's safe to assume that the Israeli pilot who knocked out the Syrian nuclear reactor cut his teeth at this game.

These games, while seemingly simple, represent the creativity of Israelis to make do with what they've got. Who knows, perhaps today's ten-year-old inventor of the next popular recess game will be the creator of tomorrow's new life-saving medical device or cybersecurity system.

How do you say "necessity is the mother of invention" in Hebrew?

Bagrut: "Will This Be on the Test?"

Hang around Israeli teenagers long enough, and you'll pick up a few words that open a window into their emotional state.

For starters, there's *giyus* (the draft). Whether they're dreaming of serving in the Paratroopers or concocting a scheme to avoid it altogether, mandatory conscription for all eighteen-year-olds weighs heavily on their minds.

Another is *"test"* – specifically the driving test, which they can take at age seventeen and a half to get their license.

And then there's *bagrut* (short for *te'udat bagrut*), the official, countrywide high school matriculation exams taken in eleventh and twelfth grades. You don't need to pass the *bagrut* to graduate – a *te'udat bagrut* (matriculation certificate) isn't the same as a diploma. But in order to attend college or serve in higher units of the IDF, including the coveted 8200 (a gateway to high-tech riches), a *bagrut meleiah* (full *bagrut*) is mandatory.

Similar to the British A-Levels and New York State Regents, *bagrut* tests are cumulative, assessing mastery of subject matter throughout the entirety of high school. Unlike in the United Kingdom and New York, however, Israeli students begin the *bagrut* in eleventh grade. Until 2020, when the school system was thrown into disarray by the COVID-19 pandemic, students were tested in each of the eight subjects set forth by the Ministry of Education: math, English, civics, literature, history, language arts, Bible, and at least one elective in the hard or soft sciences, a foreign language, or the arts.

In state-run religious Jewish schools, students also had to pass Talmud; in state-run Arab institutions, Islamic, Druze, or Christian heritage instead of the Hebrew Bible, and Arabic in addition to Hebrew. With this many required subject matters for *bagrut meleiah*, Israel's was the highest number of matriculation tests of any country in the world.

It's not only students who are losing sleep; *bagrut* tests are stressful for teachers, too. Because no matter how great they are in the classroom, both the teachers and school are ultimately measured by students' test scores. Not surprisingly, this often leads to "teaching to the test," especially in the final weeks before an impending exam. It's not uncommon for students to be called back to school during the first week of Passover vacation for "optional" (but really mandatory) marathon sessions where teachers cram in any and all remaining material for the early May exams, especially in math and science. So much for the Holiday of Freedom.

Bagrut scores are weighted to account for difficulty, the same way a B+ in Advanced Placement English in the US counts more toward a GPA (grade point average) than an A in the regular track, so that a 75 percent in a Level 5 subject (the most difficult) results in a higher score than an 85 percent in Level 4 or even a 99 percent in the lowest, Level 3.

To prevent cheating, the exam for each subject is administered on the same date at the same time for every student in the country, with the same exact test. The physical tests themselves are therefore treated like gold, delivered to schools at the last possible moment and kept under lock and key until it's go time. Tests are administered by independent proctors-for-hire who have no stake in the outcome of the tests (unlike teachers), and are graded off-site by current and former teachers who don't know the identity of the students.

Still, every so often a test will somehow get out and make the rounds on messaging apps or social media. When this happens, a wrench is thrown into the entire education system, forcing officials to reschedule the test with a fresh exam or simply accept a "contaminated" set of scores. In one such case in 2022, a group on the messaging app Telegram sold answers to the math test for the low price of 300 shekels. Unbeknownst to them, group members included representatives from the Ministry of Education, quietly watching the correspondence with the knowledge that, for the very first time, students would receive one of *three* different versions of the test. Sample question: If Uri has 300 shekels and flushes it down the toilet on answers to a test he won't be taking, how many shekels does Uri have left?

In early 2022, Education Minister Yifat Shasha-Biton announced her intention to replace several of the *bagrut* tests (history, literature, civics, and Bible) with research projects and other assignments.

Whether these reforms will lead to a brighter future for Israeli students is unknown. What is known is that the many recently proposed changes are but one more outcome of government instability. As Dr. Michal Shaul of Herzog College said, "Our teachers, worn out by two years of the COVID-19 pandemic and exhausted by the frequent reforms imposed on them, now have to deal with another reformer education minister and her desire to make her mark. As soon as one new reform has been internalized by our schools, another reform comes along to shake things up again."[3]

Perhaps the biggest question of all, however, is whether high school students are learning what they need to in order to be vibrant,

---

3. Michal Shaul, "Are Israeli High School Graduates Ready for Assessment?," *Jerusalem Post*, February 12, 2022.

productive members of society ... or merely what needs to be learned to pass the *bagrut*.

Though all countries face this question, in Israel's case, the answer may just lie in the test itself. Almost by definition, any standardized national test is itself a reflection of that society's priorities. "Twenty years ago, a student could take a Level 5 test in many areas of the humanities, like philosophy or literature," said Shadmit, a high school teacher in Herzliya. "Today, Level 5 is offered only in the sciences and English, because that's what our government thinks is most important to Israel's future.

"I get it," Shadmit says. "But what about kids who don't want to go into high tech? If you happen to love literature or theater, or if you want to explore philosophy, where do you fit in? Does Israel have a place for you?"

Darca Druze High: A School Built on Values

"Being Druze, it's complicated," says Ali, an eleventh-grader at the Darca Druze High School for Science and Leadership in Yarka.

I ask Ali to explain. He hesitates, then glances across the desk at his principal, Kamil Shela, who nods.

"Basically," Ali says, "the Arabs think we're traitors, since we serve in the Israeli army."

"And the Jews?" I ask.

"That we're terrorists. Because we're Arabs."

Many Jewish Israelis interact with Druze primarily on weekends, when they pull to the side of the highway for some *pita Druzit* – homemade Druze pita, kneaded, flattened, and cooked on an open-flame stove called a *saj* on the spot and then filled with traditional *za'atar* and *labaneh* (yogurt) or, less traditionally, Nutella. Druze towns like Daliyat al-Carmel are a popular Sabbath destination for their authentic Arab food and outdoor markets.

But there is, obviously, far more to the Druze than great shopping and tasty snacks. Derived from Islam a thousand years ago, Druze is a monotheistic religion that, Shela explains, also draws inspiration from Christianity, Judaism, Hinduism, and even the teachings of Socrates and Plato. They recognize seven prophets, the holiest being Jethro, father-in-law of Moses, and others including Abraham, Jesus, and Mohammed. There are a million Druze in the Middle East today, almost all in Syria and Lebanon but about 150,000 in Israel's Galilee and Golan Heights. Of the many unique aspects of their religion, one is that they are loyal to the government in power, whoever that may be. Unlike Muslim Arabs, Druze are drafted into the IDF, often into the Border Guard and other combat units.

Another core precept of the Druze is that they believe in reincarnation, or as Shela words it, "transformation of souls." At the moment of death, a Druze person's soul is transferred to the body of a newborn of the same gender, immediately. Nor is there conversion into or out of the Druze people. "What this means," Shela says, "is that the number of Druze souls in the universe is and always will be fixed."

Until 2011, Shela's high school in Yarka, a Druze village in the Galilee, was like most Arab-Israeli schools: underfunded and underperforming. That year, the Ministry of Education and US-based Rashi Foundation cocreated Darca, a network of subsidized schools to help level the playing field among the country's underserved communities.

Seven schools took part in the Darca program during its pilot year; today, there are forty-three. The Darca mission is pursued through two strategies: funding – for everything from computer labs to faculty – and leadership.

That's where Kamil Shela comes in. A winner of the National Educator of the Year award, Shela took over Yarka in 2014. That year, and every year since, the Darca Druze High School for Science and Leadership has been number one in the country.

How can it be that the top high school in Israel, homeland of the People of the Book, isn't actually Jewish, but Arab Druze? But the numbers don't lie: since 2014, more than 60 percent of Darca Druze students score 90 or higher on the *bagrut* matriculation exam on Level 5 math and English and easily exceed the required number of volunteer hours. At the other nine schools in the top ten, about 40 percent of students score 90 or above. (In any given year, about 50 percent of Israeli students pass the *bagrut* at all.) Similarly, before 2014, almost no Yarka graduates went to university. Today, more than 70 percent qualify.

Ali is one of them. After graduation, he'll enlist in a special academic-military track where he'll simultaneously complete his mandatory IDF service and attend medical school.

When I ask Shela the secret to the school's success, he doesn't hesitate. "*Arachim*," he says in Hebrew: values. "I could talk about test scores and pedagogical theory, but the foundation is and always will be values, foremost among them respect for one's parents. Respecting elders is one of the most important values to the Druze."

Shela credits two other ideas for the school's success. The first is faculty: before teachers are hired, they must teach a sample class while being observed and rated by Shela, the department head, the regular teacher, and students.

The other is what's known in Hebrew as *avirah* (atmosphere). "A few years ago, we won the National Prize for Education. I asked the head of the committee why she chose us from among so many deserving schools. She said everywhere else, schools were afraid that extracurricular activities and social programs would interfere with academics. We were the only ones afraid that academics would interfere with activities, social programs, and volunteering."

The three students who meet me, Ali and two twelfth-grade young women named Yana and Samar, all gush about their beloved school; the teachers, who look after them like family; and their fellow students, whom they refer to as brothers and sisters. Not even a long commute – for some, up to ninety minutes each way – dispirits them. "At my old school, my teachers didn't necessarily believe I'd amount to anything," says Samar about the school she attended in the Druze village of Usafiya. "Here, all I hear is encouragement." Next year, Samar plans to do Sherut Leumi (National Service), an alternative to serving in the IDF, and then attend the Technion.

When I ask what it's like to be Druze in a Jewish country, all three tell me they've experienced racism – despite their high academic achievements and willingness to put their lives on the line for the country.

Yana brings up the 2018 Chok Leumi (Nationalist Law), which declared, among other things, that Israel is the nation-state of the Jews, with Hebrew as the official language. (Arabic was granted "special status.") "It's like we're not wanted here," Yana says. "We're allowed to live here, but we don't really belong."

Shela himself has criticized the law from the outset. After he decried it in several media appearances, some Druze leaders suggested he run for the Knesset – an offer he turned down.

When I ask why, Shela leans forward on his desk. "Teachers can change the world," he says. "But politicians?" He thinks for a moment. "Politics and values are two things that just don't mix."

CHAPTER 5

The Economy, Work, and Work-Life Balance

STOP ME IF YOU'VE HEARD THIS ONE:

How do you make a small fortune in Israel?

Start with a large one.

How old is that joke? So old that immigrants from the early days of socialism, hyperinflation, and the pre-shekel lira might not believe the standard of living that many Israelis enjoy today. Of course, it doesn't hurt if you can sell a navigation app to Google for a billion dollars or play Wonder Woman in a blockbuster movie.

If all you knew about Israel came from the news, you'd swear nothing happened here but war, elections, and start-ups that sell to Google for a billion dollars.

Believe it or not, there's a lot more to Israel than what makes it into the headlines. Day-to-day life in Israel is more or less like life anywhere: Grown-ups go to work. Families take trips to the beach and hang out in the park. People meet, date, and fall in love, sometimes with more than one person (we'll explain).

Still, ask Israelis what daily life is like, and they'll probably answer "hard." Not because of war or conflict, but economics. In 2011, a million people took to the streets to protest the cost of everything from rent to cottage cheese. The struggle to *soger et hachodesh* without any debt is a national pastime; fortunately, overdraft is not only allowed in Israel, the banks encourage it. Making it even harder is that when they look across the Mediterranean to their European neighbors, Israelis see a cost of living that is 30 percent lower in France and 40 percent lower in Spain.

What's Hebrew for "*Ay, caramba!*"?

In chapter 5, we'll explore the Israeli economy, including how people manage to get by (or don't) in a country where salaries are low, taxes are high, and import tariffs on cars are as much as the cars themselves. We'll also visit the Israeli office to explore corporate culture (no need to dress up), unpack some of the trends and jobs unique to the Israeli labor market, and learn how Israelis manage to balance all this with perhaps an even harder job: parenting.

And if any of your Israeli friends ask to borrow this book instead of buying their own copy, don't be surprised. Israelis will do anything to save a few *grushim* (slang for the tiniest possible currency, "pennies"), even when they don't necessarily have to.

How else to make a small fortune last?

Tama 38: A Benevolent Plan on Shaky Ground

When you travel Israel by car, you can't help but notice how drastically and quickly the landscape changes. It's one of the most magical aspects of the country – the transition from rolling green hills to urban steel and concrete, from pastoral farmland and forest to desert, all in under an hour.

But pay attention as you drive this beautiful country, and you're bound to notice something else: cranes. Tall, towering, hundred-meter-high construction cranes, looming over sand pits and scaffold-wrapped apartment buildings, often with a cable attached to lift and place piping and concrete slabs. Pick a town, any town, and we guarantee you that within three minutes, you'll see a crane, usually with a dump truck nearby and a dozen or more workers milling about.

"Medinah mitpatachat, mah la'asot?" Israelis are wont to say. "We're a young country, still being built – what are you going to do?"

And it's true. With a population that's expected to hit fifteen million by 2050, Israel is frantically trying to build enough housing. And since Israel is small (about 8,600 square miles, including the West Bank), the only choice is to build *up*.

Hence the cranes.

But there's another explanation for the ubiquitous cement mixers, cranes, and piles of rubble. One that originates long before Israel became an independent state – four billion years before, to be exact.

Just below the eastern edge of the country lies the Syrian-African Rift, a fault line along the Jordan River that makes Israel prone – *extremely* prone – to earthquakes (as if we didn't have enough problems already). The last major quake was in 1927 and measured 6.2 on the Richter scale, causing extensive damage and leaving more than five hundred dead in Mandatory Palestine and Transjordan. According to seismologists, the next big earthquake is not a matter of *if*, but *when*.

Most apartment buildings and houses, however, are not strong enough to withstand an earthquake. When the country's first residences were built, in the early decades of the country, the main goal was to erect housing as quickly as possible, protective reinforcements be damned. As a result, this early construction is a disaster waiting to happen.

To rectify this, in the early years of the new millennium, the Israeli government initiated a plan called Tama 38. Tama is a Hebrew abbreviation for "National Outline Plan," and 38 is an administrative code. Today, "Tama" is a term that Israelis all know and actively disdain, unless they work in construction or real estate.

"The logic of Tama 38 was so simple," says Tomer Berreby, a real estate lawyer in Tel Aviv. "The cost of fortifying hundreds of thousands of apartment buildings is prohibitive, so the government worked out a deal: Contractors fortify all pre-1980 apartment buildings at their own expense, and in exchange, they get to add two stories to these buildings, which they can divide into individual apartments and sell for profit."

To compensate residents, contractors upgrade the exterior of the building, add an elevator (since most old buildings don't have one), increase the size of individual apartments by adding a room or balcony, and provide other perks that drive up the value of each apartment by more than 20 percent, with no financial cost incurred by the tenants.

"On paper, Tama is a win-win-win," Berreby says. "The government protects millions of its citizens against an impending natural disaster at no cost. Apartment owners see their homes go up in value. And contractors reap tons of cash."

What's not to like?

Well, jackhammers, for one thing.

"I'm basically living on a construction site," says Shani, a thirty-nine-year-old married mother of two in Ra'anana. Though Shani's building isn't undergoing Tama, the one next door has been – for nearly five years. "Six days a week, at seven a.m. sharp, it begins…jackhammers, drilling, hammering, welding. Not to mention the crew's nonstop shouting and the cigarette butts littering our sidewalk. I can't use my beautiful porch because it's covered in dust. And I just found out that the building on the other side of me is starting Tama next month. It's hell."

Shani is far from alone in her disgust. Residents who live near a Tama site – which, in the center of the country, is just about everywhere – experience a dramatic reduction in their quality of life.

Still, one could argue that no matter how noisy, dirty, and disruptive it is, Tama is worth it for one simple reason: it saves lives.

Unless, of course, it doesn't.

"The politicians who initiated Tama meant well," says the Technion's Dr. Nir Mualam, an associate professor in the faculty of architecture and town planning. The problem, Mualam explains, is that almost all Tama fortification projects have been implemented in Tel Aviv, Ra'anana, Herzliya, and other wealthy cities where contractors can earn huge profits from selling new apartments, but barely at all in Tiberias, Eilat, and other areas along the Syrian-African Rift. In other words, the cities that are most vulnerable to ruin aren't being fortified because the profit margins are too low.

"Our leaders miscalculated when they decided to make Tama a free market project," Mualam says. "It's led to a complete diversion of goals, from earthquake fortification to monetary profit. Tel Aviv and Herzliya get redone, but Beit Shean and Eilat don't, because land prices there aren't as high and contractors have less to gain."

Berreby agrees. "If, God forbid, the big one hits, and all these newly fortified buildings all remain standing, we'll look back on Tama as one of the ingenious projects in Israel's history.

"For now, though," he says, "most folks involved are much less concerned with fortifying buildings against earthquakes than they are about fortifying something else: their wallets."

Face to Face with Real Estate Expert Dr. Danny Ben-Shahar

While Israeli folk songs like "Shir l'Shalom" (A Song for Peace) and "Yihyeh Tov" (It Will Be Good) dream of peace with our neighbors, Israelis' minds are more often filled with another, more tangible dream: home ownership. With Tel Aviv named one of the world's most expensive cities and a one-bedroom Jerusalem apartment costing well over a million dollars, this dream is on life support, with the classic path of "army → university → buying a starter home" no longer viable. How did this happen, and what economic future does the market hold?

To understand the evolution of the local housing market and the potential solutions to making ownership more affordable for all, we spoke to Dr. Danny Ben-Shahar, professor of finance and real estate at the Coller School of Management, Tel Aviv University, and director of the Alrov Institute for Real Estate Research.

Joel Chasnoff and Benji Lovitt: When we lived in the States, most of our middle-class friends bought homes in their twenties or thirties. What is so different about the real estate market in Israel?

Danny Ben-Shahar: There are many factors, but the fundamental issue is that more than 90 percent of the land in Israel is owned by the government, through a body called the Israel Land Authority, and the government further monopolizes all planning processes. This makes construction much more difficult.

How exactly?

When you want to build a home or apartment building in the US, you can work directly with a contractor or real estate developer, arrange financing with a bank, and completely design and build the house in about a year. Here, you need to wait for the government to release the land and the committees to approve your construction plan and issue the permit. That means bureaucracy, committees, a lot of red tape… so for a number of years or more, nothing happens.

Especially when the government is unstable and can't approve a new budget for years.

Exactly. The entire process – what you can build, how big, how much – is decided by political officials. From the time you find a plot, you could be waiting for years just to break ground.

How does that lead to such high housing costs?

It's an issue of supply and demand. Because the government controls the supply, and because they are so slow in releasing land and approving the start of new housing developments, there simply aren't enough homes to purchase for a population that's growing at an annual rate of about 1.8 percent – the highest among OECD [Organization for Economic Cooperation and Development] countries. What does exist is very uniform. But there's another factor too: the vast majority of homes in the country are in privately owned apartment buildings, by which I mean that each apartment within the building belongs to someone else. Multi-family projects – big complexes with a central management company and a pool, gym, and other amenities – are only starting to exist. Without the economies of scale, it not only makes everything more expensive, it considerably slows attempts for urban renewal and gentrification.

Salaries are low and taxes are high. How much can someone expect to pay for an apartment?

As of the second quarter of 2022, nationally, the average three-bedroom unit costs about $760,000 [$1 equals about NIS 3.5]. Under current regulations, that requires a down payment of no less than $190,000. And this is with an average household net income of about $6,200 per month.

So when people can barely finish the month with money in their account, how do they do it?

These days, most don't. That's why we had a million people protesting in the streets in 2011, living in tents. In fact, things have gotten worse over the last decade, not better.

And for those who do manage to buy?

Those who do buy do a couple of things. It's very common for young couples to receive help from their parents. That's part of the culture here. Others have to take out loans, but not the typical kind of mortgage you're thinking about. Because homeowners have to pay at least 25 percent down, many people take out a loan for that.

Which means they're taking out a loan just for the right to take out a loan?

Now you're getting it! This motivates middle-class people to consider leaving the country. It's just not sustainable.

What needs to change to make real estate more affordable?

Number one, the government needs to release more land, especially in the north and south of the country. This tiny country only has one true metropolitan city, around which the best economic opportunities and jobs exist, not to mention much of its culture. When so many people want to live in the Merkaz (central Israel), the prices skyrocket. We say in Hebrew, most of the country lives in the area defined as "from Hadera to Gedera." It's a nice rhyme, but we need better transportation infrastructure and high-speed trains to make living in the outer areas, what we call the *peripheriah* (periphery), more appealing. Number two, the entire process has to move faster. Israel is experiencing the highest population growth rate among the entire OECD. This means that within the next four decades, we need to double our current number of housing units to keep up. We have 2.6 million housing units, but we will need more than five. Lastly, we need to do a better job of assisting families in need. Only 1 percent of housing in Israel is low income. I'd like to see more public housing and rent subsidies for the lower income deciles as well as affordable housing programs, or we're going to see a homeless problem that only gets worse.

Final question: If we had a million shekels to invest in real estate anywhere in Israel, where would that be?

[*Laughs.*] I don't think a million shekels would be nearly enough. But I'll tell you this – if I knew for sure, I'd be a billionaire.

Shopping: "Mr. Zol" as a Way of Life

True or false: Israelis would purchase a one-way ticket to Syria just to buy a PlayStation at the airport duty free.

The answer: false, but only because the TLV–DAM route hasn't opened up yet.

Israel is crazy expensive. High import taxes, high defense budget, and low salaries are just three of the factors that contribute to create an economic environment where millions of people struggle just to pay the bills. Yes, Israel is the start-up nation and home to a growing number of tech millionaires (and billionaires), and you see plenty of Teslas and even the occasional Ferrari cruising the streets. Still, "around 90 percent of Israelis do not work in the tech sector and lack access to the higher salaries of high tech," says Yoav Fisher, a Tel Aviv–based economist. "The increasing wage disparities between sectors create a noticeable and persistent challenge for the 90 percent to *soger et hachodesh* without any debts."

So how do Israelis survive? Fisher explains, "There are important aspects of Israeli society that are nationalized and supported by the government." The safety net embedded in the country's socialist roots provides healthcare for all (eliminating the need for American-style GoFundMe campaigns for medical procedures); it also grants affordable higher education, generous pension plans, and a strong welfare system that keeps people from living in the streets. Only rarely will one see homeless people in Israel, and never children.

But there's another way that Israelis overcome high prices and low salaries to stay afloat: by obsessively pursuing bargains wherever they can. In fact, whereas "Jews are cheap" is considered an insult abroad, Israelis wear the stereotype as a badge of honor, which explains why they won't hesitate to tell you about the great deal they just got on their new fridge or high-speed internet service. Sellers know it too and advertise in kind. Take, for example, grocery stores Chatzi Chinam (Half-Free) and Mr. Zol (Mr. Cheap). Do they know their customers, or what? In Israel, "saving money" isn't about contributing to a child's college fund; it's about simply making ends meet.

That said, a deal here or there isn't enough to get by. To *really* save money, Israelis must be constantly vigilant and creative.

Here are a few methods Israelis employ to save the proverbial shekel or two.

Find a mule. If you visit Israel frequently, you've probably been asked to fill your suitcase with half of Target or, more recently, Amazon. This tradition dates back decades, though needs have changed. In the seventies, Israelis asked visitors to bring items that were either hard to find or literally unavailable: Colgate toothpaste, Guess jeans, Secret antiperspirant, and Oreos. Today, when your friend Michal asks you to bring her favorite shampoo that smells like coconut, it's not because local stores don't sell hair products but because that same item will cost her three times as much locally. Same thing with iPhones, Google Chromecasts, and USB chargers, which just underscores the importance of tourism: when people stopped coming during the Second

Intifada and COVID-19, not only did it nearly wreck the local economy, but Israeli parents suddenly had to pay full price for Huggies.

Get a lot in Eilat. Ahhh, Eilat: the southern resort of sun, snorkeling… and a more affordable Xbox. Hey, you can get a tan anywhere, but only in Eilat can you purchase major appliances minus the nationwide 17 percent value added tax. You're allowed one telephone and one laptop per person every six months, tracked through your *te'udat zehut* (personal ID) number; however, if you bring your family of five, you can walk out with one for each (contingent on proof that you're actually related – a printout from the Ministry of the Interior will suffice, though the wait for it will likely be longer than the round-trip drive to and from Eilat). There's a calculus involved, of course; factor in gas or bus, lodging for a night, meals on the road, and you might decide to just buy your iPad at the local mall. But if you do go to Eilat, when is the best time to visit? Black Friday! (Double the discount! Come on, people, time to start thinking like a local…)

Fly 'n buy. Another popular money-saving strategy is to simply leave the country. Israelis have been known to combine shopping with a family vacation in London, Paris, and other high-end destinations. As of this writing, there's still no Harrods in Israel, despite Tel Aviv being the world's most expensive city. (Jerusalem does have the Kotel, built by Herod the Great, but it doesn't yet offer a rewards membership.) When Israeli families fly overseas, they schedule at least one full day to shop at Primark, Nike, H&M, and more, often stocking up on kids' clothing for the next two or three years. Then, and only then, will they visit the Van Gogh Museum or Prague Castle (time permitting). "But Israel *has* H&M!" you're thinking. True. But the prices are literally two or three times higher.

That explains the need for affordable internet. When you just can't wait for Aunt Jenny to schlep over electronics, makeup, and soccer ball pumps, prepare to spend hours on Ali Express, China's version of Amazon, a virtual online dollar store with minimal shipping costs, though their products have been known to be unreliable. (The Apple iPhone charger you ordered may show up three months later as an electric rechargeable apple peeler.) Recently, Amazon commenced shipping to Israel, but there's a catch: to justify the costs, buyers must thread the needle of buying enough items to avoid exorbitant shipping charges (over $49) but not so many as to incur customs fees (under $75). If you really want that wallet of your desire, you may find yourself adding taco seasoning, waxed dental floss, and a doggy chew toy to your cart just to hit the sweet spot. Doesn't it feel like you're on *The Price Is Right*? Whoops, a message appears that not all these vendors ship internationally. You lose, and thanks for playing.

All *shuk* up. But the most traditional way to stretch your shekels is by shopping at the *shuk*, the traditional Middle Eastern market. The *shuk* attracts every sort and stratum of person because it has – and we mean it – *everything*. Need a spatula? Got it. Suitcase? Check. Toenail clippers? Yep. A Paris Saint-Germain soccer jersey, size eight, with Lionel Messi's last name on the back spelled "MESI"? Right over there, between the guy selling bull's testicles (perfect in Yemenite soup, we hear from a friend) and the woman who'll write your name on a grain of rice.

Not only does it have everything, but the *shuk* is also the best place to find a bargain, offering linens, blenders, automatic trash cans, and other housewares at rock-bottom prices, as well as all manner of Judaica, footwear, and clothing. A bonus? The recent gentrification of Jerusalem and Tel Aviv's markets has imported restaurants and bars whose invigorating energy is hard to find anywhere else – especially on Fridays, in the final hours before Shabbat.

And that's when you should go. In those waning moments before the sun sets on Friday afternoon, the *shuk* is packed with people bustling about, rushing and readying for the Sabbath, just as Jews have for thousands of years. There's something very Second Temple–era about the vibe (if the Sadducees drank cappuccinos).

Because if Israel is all about the highs and lows, the bitter and the sweet, *eizeh chaval* (what a shame) to stress about the high cost of living if you're not going to also enjoy life along the way. And if you want to stop and smell the roses, you could do worse than the fresh ones sold at the *shuk*. Be sure to tell your Israeli friends how little you paid – we promise they'll be impressed.

Rescue 101: Ambulance on Demand

If, on your next visit to Israel, you should happen to have a medical emergency and need an ambulance (*chas v'chalilah* [heaven forbid!], as they say here), be sure to thank the first responders who treat you. Of the thirty-four thousand paramedics in Israel, thirty thousand are volunteers (thousands of whom are Haredi, incidentally, or teens).

While you're at it, thank the good people of Helsinki, Melbourne, and fourteen other Jewish and Christian communities around the world. They're the ones who paid for the ambulance itself.

"In this, Israel is completely unique," says Eli Jaffe, director of training and volunteers for Magen David Adom, the Israeli branch of the International Red Cross. "Everywhere else, ambulances are funded by local or state governments. Here, since we're a Red Cross organization, we can't be affiliated with the government. So we've been completely dependent on donations since day one."

Actually, even before day one. As Jaffe explains, Israel got its first ambulance in 1930, when a group of pioneers got together and raised money from fellow settlers. "Israeli civilians still contribute," he says. "But these days, almost all our contributions are from Jews and Christians overseas."

The good news is, you'll always know exactly whom to thank. On the front door of every ambulance is a dedication such as, "Presented to the People of Israel by Milton and Beverly Katz of Palm Beach, FL." So you can rest assured, even as you receive chest compressions and your life flashes before your eyes, the Katzes in southern Florida have your back.

Israel's reliance on outside funding isn't the only thing about its ambulance service that's unique. Even more impressive, in Jaffe's opinion, is what's known as *sharsheret hahisardut* (chain of survival) – an app-based notification nexus that alerts nearby first responders when help is needed.

Jaffe explains how it works. "In the rest of the world, when you call 911 or whatever emergency number, they send an ambulance and then relay instructions over the phone for what to do until that ambulance arrives.

"In Israel, we do all that, but also something else: when someone calls 101 [the emergency number in Israel], the call taker sends a notification to the phones of the five first responders who happen to be closest to the scene. These are off-duty doctors, nurses, and volunteers who, until that moment, were simply going about their daily lives, sitting in a café, at the park with their kids, at the office. Then the notification comes in, stating the location of the medical emergency." Whoever responds first goes immediately to the scene. If none of these five respond, the notification goes to the five who are next closest, and so on.

"We've saved countless lives thanks to this system," Jaffe says. "With thirty thousand volunteers, we have the entire country covered. There's always a first responder nearby, sometimes less than twenty meters away."

Another unique aspect of first responder care is the number of teenage volunteers. Every year, more than three thousand teens take the required sixty-hour volunteer course, often as a way to fulfill their high school's community service requirement. Once certified, these teens, some as young as fifteen, ride in the back of ambulances and function as a regular part of the crew. They're also available to be called through *sharsheret hahisardut*, making themselves available to help when called upon.

And also when they're not called upon, but just happen to be on the scene. Jaffe tells the story of a seventeen-year-old high school student who, in addition to volunteering as a first responder, works as a pizza delivery boy. One night, a family ordered a pizza; moments later, their baby began having a seizure. "When the delivery boy arrived," Jaffe says, "he heard screaming from the apartment. He knocked loudly, told the parents he was a volunteer medic, and sprang into action. He knew exactly what to do and wrapped the baby in a wet towel to bring its temperature down." By the time the ambulance arrived, the baby had stopped convulsing.

Emergency medical help comes at a cost, however. Israel's fleet of ambulances, helicopters, defibrillators, and other emergency equipment costs more than a hundred million shekels (about 30 million US dollars) a year.

Raising this kind of money year after year is challenging, Jaffe says, though not necessarily a hard sell. "People love that the impact is immediate. I mean, ambulances – what's more important than that?"

What also attracts people is the equity of the mission. "We treat everyone, everywhere in the country, no matter what religion they are or what language they speak." Moreover, first responder volunteers come from every segment of society.

"If I said to you, 'A Christian, a Muslim, and a Jew are in the back of an ambulance,' you'd think it was the start of a joke," says Jaffe. "But it's not. It's what our volunteers do every day, together. Saving lives."

My Son, the ~~Jewish~~ Muslim Doctor

The Jewish man leans over the counter and flirtatiously asks the female pharmacist, "Tell me, are you 'Alma' with an *aleph* or *ayin*?"

"Alma with a *samech*," she answers, before gently rejecting his advances so she can return to her customers, and apologizes that her tone is *yavesh, karir, v'mutzal* (dry, cold, and shaded – also the requirements for storing medicine).

Meet Salma, the Arab pharmacist played by actress Liat Harlev on popular sketch comedy show *Eretz Nehederet*, and the lens through which millions of viewers learn about daily life for Arab-Israelis. It's not by chance that she's a pharmacist by trade; a 2015 paper from the Taub Center reported that, despite making up just 20 percent of the population, Arab-Israelis represent exactly one third of the country's pharmacists. In fact, more than 70 percent of the pharmacists at Super-Pharm branches are Arab-Israelis. What is the root of this trend and how are the real-life "Salmas" changing the country?

With a historic 1994 handshake, Prime Minister Rabin and Jordan's King Hussein formalized a peace deal between their two countries. Talk of increased economic opportunities wasn't just lip service; with Hebrew U and Ben-Gurion University no longer the only nearby options, the number of Arab-Israeli pharmacists skyrocketed, with over a third of all licensed pharmacists since 2005 having studied in Jordan. More importantly, according to Alex Weinreb, research director of the Taub Center, this new option didn't only create more spots for Israeli students, it opened the door to an entirely new demographic: Muslim women. Suddenly, families whose traditions frowned upon single, unmarried women living alone or outside their houses were amenable to the culturally friendly and familiar environment of Jordan.

While the pharmacy business boomed, so did Israel's well-documented tech sector, but not for everyone equally. Weinreb explains: "The pathway into much of the tech world starts with the army. Artificial intelligence, security... graduates of the elite 8200 intelligence unit are the prime candidates. Arabs of the same ability and motivation have neither these options nor the *protektziah*."

For motivated, intelligent Arab-Israelis looking for a high-status, stable profession, what they do have is pharmaceuticals. But this story is about more than just the increased representation in this field. It's also about fundamental changes happening in the Arab-Israeli community. Developing social norms are allowing for greater freedom of mobility for jobs, which explains why you can see so many more Arab-Israeli women in central Israel than just a few years ago. Also, as Jewish Israelis have gained more wealth this century, so too have Arab-Israelis, creating a new middle class of young, skilled families with money to spend. This money is increasingly spent on higher education, especially among women. The Taub Center's latest data reveals that in the 2021–2022 academic year, two-thirds of all Arab students in bachelor's or PhD programs are women, as well as three-quarters of master's programs.

What happens when people with wealth and education have their family's blessing to move out? Many of them leave their enclaves in northern Israel and spread out across the country. Not only in the traditional mixed cities of Acco, Haifa, Jaffa, Ramle, and Lod, but also in communities such as Carmiel (which recently crossed the 5 percent threshold of Arab population that constitutes a "mixed city," according to Israel's Central Bureau of Statistics), as well as Be'ersheva, whose Degania primary school has seen its percentage of Arab students climb from 5 to 70 percent in just fifteen years.

To most TV viewers, the fictional Salma represents the phenomenon of Arabs entering the field of pharmaceuticals; since the character debuted in 2015, however, she has grown to represent much more, whether the audience knows it or not. The Taub Center's latest data from 2014 to 2016 shows that Arab-Israelis now represent over 30 percent of newly registered doctors, for the same reasons as pharmaceuticals: peace with Jordan, increased wealth, fewer opportunities in tech…not to mention that Muslim families have always valued a career in medicine. (Sound familiar, Jewish parents?)

Dr. Khaled Zbidat is a senior doctor in the emergency medicine department of Afula's Emek Medical Center. "The younger generations of Arabs no longer want jobs in physical labor," he explains. "In today's modern world, they want an education to work in jobs like medicine and tech." Which explains how the Arab city of Arahba boasts one of the highest percentages of physicians on the planet, with more than six doctors per thousand residents (compared to 3.4 nationwide). Zbidat too has friends who've moved to Carmiel for a better quality of life. "Where I live, in the village of Sachnin, there's a tremendous housing shortage, so young couples can't afford to purchase. The bigger cities like Carmiel have better infrastructure, entertainment, and services for children."

Anyone who's visited a Jerusalem hospital recently can attest to the large number of Arab staff: administrative, nurses, and doctors. And that's the most interesting part – unlike many workplaces where the management are Jewish and lower-level positions Arab, the hospitals have equality up and down the organizational chart. "The medical staff are disproportionately Arab and *kippah srugah* [literally, "knitted kippah," connoting the group of observant Jews called "religious Zionists"]," says Weinreb, "the two populations most in opposition in Israel. In the operating room, shoulder to shoulder, covered in the blood of the same patient, working together toward a very serious common goal. How can this not strengthen the bond between these peoples?"

Whatever someone's personal politics, they leave them at the door the moment they enter the hospital. "If not, it would affect your judgment, relationships, and the health of the patient you're treating," says Zbidat. "That's the whole point of the Hippocratic Oath."

If conflict can drive us apart by bringing out our worst impulses, the pandemic brought us together by forcing many Jewish Israelis to appreciate the contributions of their Arab compatriots. Not only did Arab nurses administer many of the shots, but Israel's COVID czar since July 2021 has been Dr. Salman Zarka, a Druze physician who serves as director of Ziv Medical Center in Tzfat. Whether it comes from the hospitals, increasingly mixed cities, or picking up a prescription at Super-Pharm, the further integration of Arabs and Jews has potential to dramatically reshape the country for the better – and heal, in more ways than one.

The Workplace: Check Your Org Chart at the Door

In 2015, a training slide from Intel USA went viral in Israel. Displaying a mere five bullet points and sixty-eight words, an excerpt from a guidebook titled *Working with Israelis* was shared by thousands of locals. It included the suggestions "Expect to be cut off regularly during a presentation" and "Visitors are often taken back by the tone or loudness of the discussion."

Did Israelis share this from a place of anger and offense? No! They laughed, agreed, and appreciated the acknowledgement of what not only defines Israeli work culture but has also helped make Israel one of the darlings of the tech world.

Much has been said about Israeli high tech in recent years. Dan Senor and Saul Singer wrote an entire book about it called *Start-Up Nation* in which they attribute much of this start-up success to Israeli values, especially those learned in the military. From a young age, Israelis are taught to think for themselves, take risks, and question (if not outright disrespect) authority, resulting in a culture of entrepreneurs who believe their idea will truly make the world a better place, doubters be damned.

But what does the Israeli workplace actually look like, once you're inside? Not surprisingly, the office is a microcosm of Israeli society itself.

"In a recent job interview, the first thing they asked was my age," says Mollie Searle Benoni, who made aliyah from the US in 2006 and works as the director of growth for a start-up in Herzliya. Forget the boilerplate questions such as where you see yourself in ten years (who plans so far ahead?); employers are more likely to want to get to know you, both to soften things up and to gauge whether you'll fit into the culture that start-ups are carefully trying to cultivate. Benoni says she's also been asked where she's from, why she made aliyah, and even whether or not she's married. "Employers are no longer allowed to ask about marital status, but they find creative ways to get around it – 'Do you have roommates?'"

By now, it should be clear that this first interview is in many ways symbolic of the Israeli workplace itself, especially if it was secured through *protektziah*. No need to dress up, by the way; jeans or slacks will do. And forget niceties such as exchanging business cards. If they need you, they'll message you on WhatsApp (and probably already have).

Mazal tov, you're hired! The first room you're likely to visit in your new office is the kitchen, since you'll be offered a drink the moment you arrive. Many office kitchens have not one microwave, but two: one for milk and the other meat, to accommodate employees who keep kosher. Some will even have a third for non-kosher food (enjoy your double bacon *treifburger*). And, of course, they're labeled with a sign reminding employees to *please* not mix them up.

The kitchen is vital not just for coffee and drinks, but as the meeting place for a communal lunch break. None of this antisocial "sandwich at your desk" stuff. Especially on Sunday (after Shabbat), many people will bring leftovers, eat together, and even share their food.

"Management *wants* us to eat together, another reason many tech companies pay for a 10bis or Cibus card (online food delivery)," explains Benoni. "You work better when you like your coworkers, and in high tech, the best ideas often come during lunch."

And that just may be the biggest strength of the Israeli workplace: the sense of deep, authentic communication among the staff. "Some of my closest friends are from work," says Benoni. "Formality creates distance, and Israeli culture is less formal. I could never talk as honestly with colleagues or my boss in the US, but here, people expect you to be real as your authentic self. Speaking your mind is what helps the company grow."

For this reason, many Israeli companies invest in these relationships through a *yom kef* (fun day). A *yom kef* could be a company-wide or department-specific day trip to a Dead Sea spa, a private movie screening at a theater, or even a weekend getaway to Greece. Human relations consultant Yuri Kruman says, "Not only are these outings fun, they help teams jell, which is even more important post-COVID. With more people working remotely, it's critical to carve out time for a team to bond, especially when certain decisions are best made in person."

This lack of separation between "business" and "pleasure" is one key to Israel's high-tech success. Professional communication is no different from personal, both in what employees say and to whom and how they say it. In the Israeli workplace, anything goes, and anyone can be disagreed with, no matter where they sit on the org chart.

"I can imagine that to an outsider, our staff meetings look like chaos," says Oded Sacher, a general manager at Microsoft in the company's Herzliya office. "It gets very loud, and we often cut each other off. People I supervise aren't afraid to tell me they think I'm wrong, and I argue with those above me."

Amid such chaos, can everyone actually hear each other?

"It depends what you mean by 'hear,'" Benoni says. While actual words might get lost in the back-and-forth, she notes that this kind of fast-paced ping-ponging of communication actually helps a group navigate issues in real time. And when employees have the freedom to question and even criticize those above them, new solutions are generated that otherwise might not have come about.

Will increased globalization and international business cause Israelis to adapt and behave more like Westerners? Interestingly enough, one change in their communication style was influenced by a most unexpected source: the COVID-19 pandemic.

Oded Sacher notes that ever since Microsoft's staff meetings transitioned online, discussions are much more organized, as the limitations of technology have forced people to listen more and interrupt less. "Our Teams application has a 'virtual hand raise' feature," Sacher says. "Our US-based staff used it naturally, but in Israel it led to a big cultural change, with people now 'raising their hands' about half the time."

Benoni, for one, is no longer offended or caught off guard by Israeli work culture. Which is not to say she approves of everything.

"In my last job interview, I wasn't offended by them asking how old I was," she concludes. "Only that they failed to say that I don't look my age."

Clowning Around...in Hospitals

In a men's room at the ALYN Family Hospital for disabled children, Ariel Keren pulls on a pair of oversized bowling shoes and snaps a red rubber ball onto his nose. He straightens his bowtie and suspenders. All that's left now is his trademark hat – a blue-and-yellow Tweedledum beanie with a propeller on top.

Keren, age forty, is one of Israel's most talented comedic actors.

In 2019, he played the absent-minded sultan Sheikh Habibi in the blockbuster comedy *The Mossad*. He's a regular in stage plays and TV commercials. A few months ago, he landed his most high-profile job yet – police detective Gabi Chen on the HOT network drama *Jerusalem*.

But Keren's favorite role is the one he's played twice a week since 2014, including today: Slinky, the medical clown.

"It's all about the nose," he tells me as we step into the hall. "When people see this nose – it doesn't matter if it's kids or parents or medical staff – the barriers come down. It's like a superpower. I can connect with any human being in the world."

Israel's Dream Doctors Project was founded in 2002 by Yaakov Shriki. According to Tzur Shriki, the current director of Dream Doctors and Yaakov's son, his dad was inspired to start the program by a happy accident.

"My father flew to Europe to visit a sick friend," the younger Shriki says. "When he got to the hospital, he got lost and wandered into the children's ward. He saw clowns interacting with the kids and decided, on the spot, he would create the program in Israel."

So the elder Shriki recruited his friend Jerome Arous (clown name: Doctor Goston) and petitioned the staff at Hadassah Ein Kerem Hospital in Jerusalem to take a chance. "The nurses had no idea what to do with us," says Jerome, who's still active on the medical clown circuit. "They agreed to give it a try."

Today, there are more than a hundred medical clowns at twenty-six hospitals throughout Israel. Unlike those in other countries, Israel's clowns actually assist in medical procedures, including administering IVs and prepping patients for surgery and chemotherapy. They interact with patients and their families of all backgrounds, religions, and languages.

Shira Friedlander (aka Shorty da Biggest), thirty-three and a medical clown since 2012, remembers her first day on the job: she was assigned to the EMMS Arab-Christian Hospital in Nazareth. "There I was, this married religious Jewish woman, walking into the Arab hospital," she recalls. Complicating the scenario was the already tense national mood: it was summer 2014, in the midst of Operation Tzuk Eitan, when IDF ground forces entered Gaza in response to rocket fire on southern Israel by Hamas. Scores of IDF soldiers and thousands of Gazans were killed. "I remember going into the dressing room and feeling so completely out of place, wondering if I would even be safe. As I changed out of my everyday clothes, I felt as if I were shedding layers – all the layers that Jews and Arabs have and that define us. Then I slipped into my clown suit, put on the red nose, and that was it – the walls came down."

That first day, Friedlander went into the preemie ward, a room full of anxious parents. "I walked around with a little music box, playing the softest music. The parents were so happy to see me," she says. "There was a mother sitting there, wearing a hijab and galabia. She looked so scared. I set the music box down and gave her a back rub, and she just kind of collapsed into me, giving me all the tension and fear stored inside. I looked through the glass at her baby, this tiny little thing, and thought, 'This is it. This is life. This is why we're here.'"

In recent years, the Dream Doctors Project has expanded. The program now works with the Padeh Medical Center Tiberias, which treats victims of sexual abuse. "The examination after sexual abuse is intrusive and intimate," Shriki says. "In this situation, the clown's job is to mirror the child's emotions. If the child screams at the doctor, 'This is terrible, I hate you!' the clown screams at the doctor even louder – 'This is terrible, I hate you!'" You'd think the clown would calm the child and reassure them that everything is okay, Shriki explains. "But after that kind of trauma, everything is not okay. That's why the clown is there: to validate the child's feelings and reassure them that whatever they're feeling is appropriate."

According to Shriki, the presence of medical clowns has actually helped with investigations. "Children who are accompanied by clowns are forthcoming with medical staff and with police." Just as importantly, the children encode the experience in their brains as a bit less traumatic than it otherwise would be.

Another big change to Dream Doctors – one that sets them apart from every other medical clowning organization in the world – is their new humanitarian wing. In the past decade they've sent clowns to Nepal after the 2015 earthquake, to Houston after Hurricane Irma in 2017, and to Pittsburgh after the Tree of Life shooting in 2018. In March 2022, clowns were sent to Ukraine, Poland, and Moldova to welcome and interact with refugees. "The sad truth is, Israel is a nation that knows what trauma is," Shriki says. "It's in our DNA. But it's also in our DNA to help, and to be first."

The mightiest and most macho of Israeli organizations, the IDF, has also recognized the added value of clowns. "There are now Israelis who, as part of their mandatory reserve military service, serve as clowns," Shriki says. "When IDF doctors and Search and Rescue soldiers fly to a disaster, there are two spots on every plane reserved for clowns."

To help them deal with the emotional toll, medical clowns receive counseling from social workers and attend conferences with fellow clowns. Among the biggest obstacles is working with children who, realistically, may soon die.

When I ask Friedlander how she copes with the fact that some of the kids she works with might not be around much longer, she takes a deep breath. For the first time in our conversation, her bubbly disposition dims.

"It's incredibly hard," she says. "But I tell myself the only moment that matters is now. This child, on his bed, and Shorty next to him holding his hand. In that moment, we are the entire world."

Face to Face with Online Dating Guru Yuval Katz

If you're a single female vegan looking for love, where to search for your vegan better half? Until recently, you might have kept your eyes peeled at your favorite salad joint or health store, hoping to bump into that special non-animal-eating someone. But that all changed once internet dating guru Yuval Katz and his longtime business partner, Daniel Bronitzky, created GreenDate.co.il, the number one vegan and vegetarian dating site in Israel.

Why did we seek out a conversation with Yuval? Because exploring niche dating sites opens a portal into the minds of Israelis and a side of them we otherwise might never encounter (certainly one we never learned about in Hebrew school).

Joel Chasnoff: Over the past twenty-plus years, you and your business partner, Daniel Bronitzky, have created some extremely niche dating sites. How did this whole enterprise start?

Yuval Katz: Just as the dot-com bubble was bursting in 2000, I finished my studies in business administration and information technology. The job market was awful, but I was young, single, and dating *a lot.* The first-generation dating sites were pretty standard: man seeking woman, woman seeking man. I wondered, "What else is possible?" So I created SheDate, the first lesbian dating site in Israel. There I was, in the bedroom of my religious parents' home, a *dati* kid creating a website for lesbians. Honestly, I didn't think the site would go anywhere, but I had to do *something* with my degree. Just a small amount of advertising created an overwhelming response. My first lesson was that there are entire populations looking for a certain kind of person whose needs were being completely ignored.

So you began creating other sites?

Yes, but something else happened, too. Because my business name appeared at the bottom of SheDate, others began contacting me about designing sites for *their* populations. And all these years later, that's what I'm still doing.

Tell me about GreenDate. Why do vegetarians and vegans need their own dating site?

Having grown up religious, I can completely relate to the challenge that vegetarians – and vegans especially – face. When a religious Jew goes out with someone secular, one person can eat anything while the other can only eat certain foods, off certain dishes, and in certain restaurants. But it's more than a logistical problem; it's about values and finding a like-minded partner. I know plenty of vegans who are as fervent about their beliefs as the Jews in Meah Shearim [an ultra-Orthodox neighborhood in Jerusalem] are about theirs.

Among your sites, you have a few that we might call . . . I don't want to be judgmental, but –

Shoneh [different].

Exactly. Like PolyDate for polyamorous couples, ZigZug for swingers. And GoBaby, which I don't fully understand . . .

GoBaby was the second site I created. It's a co-parenting site for strangers who want to raise a child together but don't want to be in a relationship. Imagine a divorced couple who share custody of their children and split up the responsibilities – that's GoBaby, except without the divorce. And we accommodate all types of arrangements: man and woman, woman and two men, two women and two men . . .

And there are people who want this?

Thousands. And let me tell you, it works beautifully. Think about it – a woman can always go to a sperm bank and raise a child alone, but there are so many challenges: the absence of the father, the lack of financial support. Children raised through GoBaby get emotional and financial support from both sides, not to mention another set of grandparents. The kids are raised in two separate but stable environments and are some of the healthiest children you'll ever meet. Why? Because the parents aren't worried about coupledom! There's no pressure to make the romantic side of the relationship work, no tension between the parents from whatever emotional baggage they each brought to the relationship.

And logistically, where do the babies come from?

Are you asking me how babies are made?

I mean –

Sometimes they're adopted, other times IVF from the couple or a donor.

So what do people think of these "different" sites?

SheDate and GoBaby get zero pushback. PolyDate, very little. ZigZug, which is for married swingers, does get me some angry emails, especially when the media does a piece on us. The other site that had issues is RichDate, for beautiful women and rich men.

How do you screen?

For the rich men, the monthly membership cost is 500 shekels. Considering that other sites in Israel are eighty shekels, that's high.

And the women?

We don't screen. Any woman can register for free, but then the men –

Survival of the fittest.

Prettiest. If that's what they're after.

And you got pushback on that?

Not so much pushback, more that it was a complete failure for the first two years. When it comes to relationships and sex, Israel is still quite conservative, and something about the site didn't jibe with the general attitude toward what it means to be in a couple. But once we got some TV coverage, the site grew exponentially. Women saw the piece and how much fun these women seemed to be having, going out with rich guys, and they thought, "Why not me?"

What have your twenty-plus years in the business taught you about human nature?

First, that people are people. The yearning for love is the same, no matter where you go.

Second, though the internet has given us new opportunities to meet people, what we ultimately crave is a deep and meaningful connection. We may have gone from writing long, wordy profiles on our computers to swiping photos on our phones, but the search for connection remains the same.

Last question: Is anything uniquely Israeli about all this?

On the dating side, no. We desire love and connection like everyone else.

But on the business side, absolutely. We've faced so many challenges over the past twenty years but never quit. We learned from our failures and innovated into the next thing.

There's something Israeli about the success of GoBaby, too. We tried it in the US, but it didn't take. Everywhere else in the world, people are debating whether or not it's right to bring kids into this messed-up world. But here, it's part of our nature to have kids. Even if that means raising them with a stranger you met online.

Parenting: The Kids Will Be Alright

"I'm happy to speak to you about Israeli parenting," says social worker Gaby Rurka-Hilkowitz in her office outside Tel Aviv. "You do realize, though, the phrase is an oxymoron?"

Israeli parents are known for being hands off, especially compared to their American peers. Not long ago, as Joel dined in a restaurant, three children nearby climbed up and sat, barefoot and cross-legged, on their table and played on their phones while their parents sat by, ignoring it all.

Rurka-Hilkowitz admits that Israeli parenting isn't as laid-back as it once was. "We're starting to see more helicopter parenting," she says, "though nothing like in America."

Asked for an example, she answers immediately. "The army. Parents used to drop their kids off at the *bakum* [IDF processing base] with 'Good luck, we love you, stay safe.' Today they call the Draft Board ahead of time and try to convince HR why their kid who got a 70 on their Level 3 *bagrut* should be in 8200 [the crème-de-la-crème intelligence unit and doorway to a future in high tech]."

Nor do parents have qualms about calling their kids' commanders, demanding to know why Avi was made to stand duty during dinner last night *and* breakfast this morning, or explaining that Maya shouldn't have to go on the platoon hike tonight because she's on her period.

Still, Israeli parenting is unique in plenty of ways. Here, according to Rurka-Hilkowitz, are seven reasons why.

Golden child. "When talking about family life in Israel, you absolutely have to start with the fact that children are considered golden," Rurka-Hilkowitz says. "Not just for Jews, but among Israeli-Arabs, too. All parents love their kids, but in Israel, it's another level." She notes that in Judaism, having kids is considered a mitzvah (commandment). "And then there's the memory, even if it's latent, of all the Jews we've lost over the years to persecution and tragedy," she says. "We believe that every life is precious, particularly those of our kids. It's led to a very child-centric culture where kids are brought everywhere with their parents, and parents very much want their children to be happy."

It's two a.m. – do you know where your child is? "In Israel, there's a sense that children are the communal responsibility of everyone," Rurka-Hilkowitz says. "If a child is crying on the sidewalk or in the park, people will come to ask what's wrong." At the same time, kidnapping and random violence are rare. She tells the story of a woman who went shopping and left her car door unlocked; a thief got in and stole the car, only to discover a baby in the back seat. The thief pulled over and called the police anonymously, not wanting to kidnap a child. Because parents trust their kids will be safe, they can be hands off. Teenagers don't really have curfew; they hang out with friends all night on weekends, camp out on the beach, and generally enjoy a level of independence not found in other places.

College? Probably. Maybe. Eventually. "In the US especially, so much of parenting is oriented toward getting the child into a good college," Rurka-Hilkowitz says. "Whatever that means." This mindset simply doesn't exist in Israel. "We're not in a hurry. Kids might do a year of volunteer work, then three years of the army, a year of work to save up for their big post-army trip, and then a year of travel. If they go to college, it's when they're older and ready to take their studies seriously." As a result, parents don't feel a push to funnel them into the system at a maddening pace.

My son, the hairdresser. There's an old joke about the Jewish grandmother pushing her two young grandchildren in a stroller. When a stranger asks, "Such beautiful kids, how old are they?" the grandmother replies, "The doctor is three and the lawyer is two." As Rurka-Hilkowitz explains, this obsession with career achievement isn't actually part of Israeli culture. "Most parents here just want their kids to grow up to be happy, healthy, solid members of society. If they grow up to be a lawyer or doctor, fine. But a hairdresser? That's also fine." We can attest that we rarely hear parents brag about their kids' professions, unless it arises organically in conversation.

Just wait until your sergeant gets home! "One big reason that Israeli parents are less strict is they know the army will discipline their kids for them," says Rurka-Hilkowitz. When they enlist at age eighteen, kids begin shouldering responsibility with life-and-death consequences and are held accountable for their actions. Because parents know this experience will teach their kids boundaries, they allow their children more freedom before the reality of army life takes over.

Both parents working. "It's very common here for both parents to work, at least part-time," says Rurka-Hilkowitz. The phenomenon of a parent at home full-time is rare. With both parents in the office, kids often come home to an empty house and learn to navigate the world for themselves. The realities of Israeli life force kids to be independent sooner; parents don't have the time or energy to micromanage their kids' lives.

Meet the grandparents. "Grandparents here are *very* involved in their grandkids' lives," says Rurka-Hilkowitz. "They drive carpool for school and other activities. In summer, the grandparents will sometimes look after the kids entirely." This isn't just cross-generational bonding time; the reliance on grandparents to help pick up the slack is no less important in getting by than the shopping strategies described in the "Mr. Zol" section above. "In general," she adds, "family members live close to each other and maintain stronger relationships with their relatives than, say, in the US, where they may live on opposite coasts and see each other once or twice a year." It's common, she says, for neighbors, friends, older siblings, and even babysitters as young as eleven or twelve to help parents with young kids. "The saying 'It takes a village' is very true in Israel, and literally."

At the heart of all this, Rurka-Hilkowitz explains, is Israelis' attitude toward having children in the first place. In Israel, it's simply expected that people will have kids – and if you have trouble getting pregnant, the country will help. "IVF [in vitro fertilization] costs about $20,000 in the US, but here you get five rounds a year for free," Rurka-Hilkowitz says. She points out that this applies not only to married couples, but to everyone. "Single women have kids. Same-sex couples have kids, often through surrogacy. Arab Israelis get the same medical and IVF benefits. It's very common for Jewish and Arab women to room together in maternity wards, where visiting families swap birth stories, gaze lovingly at each other's newborns, and, of course, share food."

She adds that in Israel, the social pressure to have children is strong, and that this can sometimes make for uncomfortable, if not unexpected, interactions. "If you meet strangers and get into a conversation and you mention you don't yet have kids," Rurka-Hilkowitz says, "don't be surprised if they ask, '*Nu?*'"

CHAPTER 6

The Israel Defense Forces

IN ISRAELI BOOKSTORES, YOU CAN FIND GREETING CARDS FOR the usual occasions – birthdays, anniversaries, bar and bat mitzvahs. But there's also an entire section called *"giyus na'im"* (have a pleasant draft) – for that special eighteen-year-old in your life who's about to join the Israel Defense Forces.

Even before your first visit to Israel, there's a good chance you have an opinion of the Israeli army. Many feel pride in a Jewish military that, after thousands of years of victimhood, allows Jews to stand up for themselves. Some support and pray for the army based on relationships, personal connections to people who have served. And still others have negative feelings toward the IDF for their perception of the army's role in the ongoing conflict.

For Israelis, though, the IDF is much more than something to opine about. It's something they live and experience: the heartbeat of the nation, the foundation upon which their society is built, not only for security but also as a network so large that it touches almost everyone.

Somewhere around their seventeenth birthday, Jewish and Druze Israelis receive a draft notice in the mail called a Tzav Rishon (First Summons), ordering them to their local IDF induction center to begin the process that will impact the rest of their lives. The American middle- and upper-class question of "Where did you go to college?"

is replaced by "Where did you serve?" Instead of the hallmates in your freshman dorm, those with whom you survived basic training or an officers' course become your lifelong friends. And while an Ivy League diploma is nothing to sneeze at, a cybersecurity company is more likely to hire graduates of Unit 8200, the IDF's famed intelligence division that serves as a breeding ground for many of Israel's success stories today.

In chapter 6, we explore the ins and outs of the legendary IDF, from the Nobel Prize winner who created its screening methodology to the lone soldiers who do what many Israelis find unthinkable: serve voluntarily.

If you're a *rosh gadol*, you'll finish this section speaking fluent army slang. And if not? Hey, there's nothing wrong with being a *chapash*.

Don't worry, we'll explain it all.

Drafting, Fast and Slow

Do you have many friends, or a few?
Do you play sports, and if so, how often and which?
Do you drink? Smoke pot? Be honest, don't just tell us what
you think we want to hear…

Most of us never have to answer questions like these. But almost all Israelis do, during their first visit to the *lishkat giyus* – the IDF induction center – where sixteen- and seventeen-year-old prospective soldiers are assessed for intelligence, motivation, and fitness for combat.

While their first visit to the induction center is a milestone both exciting and daunting, it's no less formidable for the military. For the IDF, the military draft resembles an HR director's nightmare: take fifty thousand job applicants, spend an hour with each, and then somehow assign them to a role that best matches their skill sets with the army's needs. It's a huge, complicated task that once again begs the question, how did anyone on this planet do anything before computers? Whoever could execute this gargantuan undertaking would surely deserve a Nobel Prize.

As it turns out, he surely did. Much of the IDF's pre-induction screening system was developed by a young officer known in his army days as Lieutenant Daniel. Today, he's known as Dr. Daniel Kahneman, winner of the 2002 Nobel Memorial Prize in economics.

As Kahneman writes in his best-selling book *Thinking, Fast and Slow*, his goal in designing a better system was to fix a glaring problem: when evaluating a young recruit's fitness for combat, interviewees who were themselves just eighteen and nineteen years old were basing their assessments on intuition instead of empirical data. Until Lieutenant Daniel discovered a better way:

> I made up a list of six characteristics that appeared relevant to per-
> formance in a combat unit, including "responsibility," "sociability,"
> and "masculine pride." I then composed, for each trait, a series
> of factual questions about the individual's life.... The idea was to
> evaluate [each recruit] as objectively as possible.[1]

Kahneman's system is largely still in use because it allows the military to do something incredible: measure the unmeasurable. "It's easy to count how many push-ups someone can do or how fast they can run," said Lieutenant Colonel C., a high-ranking officer in the Human Resources Division. "But how do you measure motivation, the ability to handle disappointment, or how well you get along with others? Kahneman's system measures those, too."

According to C., the IDF's primary assessment goal has never changed: to find the cream of the crop, best suited for the elite combat units (fighter pilots, Navy SEALS, and Special Forces) and intelligence units like the storied 8200. Before Kahneman, the military relied only on a candidate's intelligence and fitness scores without considering social and emotional traits. Many young soldiers enlisted in elite units only to drop out a few weeks later.

All told, the IDF assesses prospective soldiers according to three

---

1. Daniel Kahneman, *Thinking, Fast and Slow* (New York: Farrar, Strauss, and Giroux, 2011), 230.

other metrics: a physical fitness score called physical fitness (Profile), intelligence (Dapar), and psychological fitness (Kabah). The top candidates are invited to further interviews and, in some cases, physical tests called *gibbushim*, to determine who will advance to the elite units. From all remaining candidates, the IDF fills its ranks.

Does the system work?

In the words of Lieutenant Colonel C.: "Not as well as it could, but as well as it can." Military talk if there ever was. But it makes sense. Given the enormous task of placing fifty thousand teenagers, a 100 percent success rate is unrealistic. Sometimes kids unfit for combat are sent to combat units, and occasionally, candidates *noflim bein hakisa'ot* (fall between the chairs), like the straight-A student assigned to an unchallenging admin job, or the outgoing, gregarious future actress who'd make a wonderful classroom instructor but is somehow assigned to logistics.

More disturbingly, claims have arisen over the years that the assessment metrics are at best biased and at worst racist, favoring wealthy, light-skinned Ashkenazi kids over economically disadvantaged, dark-skinned Sepharadi, Mizrachi, and Ethiopian kids. At first glance, the claims have some merit; a disproportionate percentage of soldiers and officers in the top units are of Ashkenazi descent.

But as C. counters, it's a case of correlation, not causation. "There's no question that wealthier Ashkenazi kids from Kfar Saba and North Tel Aviv tend to score higher on the Kabah than kids from Dimona and Yerucham. But that's because kids from outside central Israel are generally less motivated to serve in elite units and come from schools that aren't as good." They're also less able to afford expensive pre-draft training courses and often face family hardships that burden commitment to an elite unit, such as coming from a single-parent home that might require frequently leaving base to help support the family. "But I assure you," C. says, "any young Israeli who is motivated to serve in an elite unit and has the intelligence and physical fitness to do so, we want them."

What about the teens themselves? To what extent does the military account for what *they* want, and what happens when they don't get it?

As Lieutenant Colonel C. explains, at the end of the day, the army's needs always come first. Not everyone can become a fighter pilot. The process can be a harsh wake-up call, not unlike how the high school valedictorian finds herself competing against hundreds of other valedictorians for a spot at Harvard.

But according to C., an even bigger challenge these days is not the kids, but parents. "The kids understand. But try explaining to a mother that her 'genius' is actually slightly above average when compared to his peers."

Face to Face with Pre-Army Fitness Trainer Daniel Alkobi

In the United States, many high school students take a course to help prepare for their university admissions exam. Israelis have recently begun enrolling in similar courses, with one difference: the exam they're preparing for is the *gibbush*, the grueling IDF pre-induction test that future soldiers need to pass to secure a spot in Shayetet-13 (Israel's Navy SEALS), Paratroopers Brigade, Air Force Fighter Pilot Training, and other elite units. One of the oldest and most respected *kosher kravi* (military fitness) courses is Xpert, co-founded by former combat soldiers Daniel Alkobi and Amit Leopold in 2012. We sat down with Alkobi to talk about the genesis of their business, how teens have changed, and the IDF's role in Israeli society.

Joel Chasnoff: How did Xpert Kosher Kravi start, and what made you and Amit think a business like this could succeed?

Daniel Alkobi: For as long as there's been an IDF, Israeli kids have trained for spots in the top units. In the past, they'd do push-ups and jog with a backpack full of books. Courses like ours actually prepare them for the conditions of the *gibbush* and, if they pass, their military service later on. With thousands of teens who dream of becoming fighter pilots, SEALS, or soldiers in other top units, we never doubted that we'd find an audience.

How has the business changed since you started?

Our first year, we had about twenty-five students. Today, we have twenty groups nationwide, each with between forty and fifty participants. When we began, the course leader basically made students run up and down hills with a sandbag. Our methods today are much more sophisticated and science-based, with participants divided into smaller cohorts according to ability, where they work specific muscle groups to enhance strength and avoid injury. Moreover, we also now address the mental demands – I always tell students that as physically grueling as the *gibbush* is, at least 30 percent of it is mental. Lastly, whereas we once aimed to help kids earn a spot in elite units, we now go one step further: to help them survive once they're in.

And by "survive," you mean...

In the elite units, two-thirds of those who start the training either quit or are booted out.

Another positive change is that, thanks to the many new combat jobs open to women, we now have a lot of girls who sign up. At first glance some of them look like what we call *banot kenyon* [mall girls], with their nails done and hair in braids. Then they jump in the mud and carry sandbags, and it's clear they're no different from the boys. Anything is possible with a solid base and motivation.

So what does a typical session include?

Wind sprints, crawling, running up and down hills with sandbags, and other physical training in the actual *gibbush*. But also plenty of mental exercises that take students by surprise.

For example?

One of my favorites is the *alumkah sociometri*, the "sociometric stretcher." While the kids stand in formation, I set up a military stretcher a hundred meters away. At the sound of my whistle, they sprint to the

stretcher; the first four to arrive have to carry one of the other participants on the stretcher for as long as they can, at a fast pace, while everyone else runs behind them, ready to swap out whoever's carrying.
The exercise is a competition to see who can suffer the most. I notice
everything: Who carries the stretcher, and for how long? Who lies
down on the stretcher? There's also the *hartza'ah spontanit* [sponta-
neous lecture]. After running and crawling to the point of exhaustion,
they're given two minutes to prepare a sixty-second lecture on the
topic of their choice to then deliver to the group. The goal is to test
their functioning while exhausted and under stress.

Who are your students, and what's their background?

I wish we drew from a cross section of Israeli society, but almost all
come from affluent, middle-class families. Not only because they can
afford the course – we do offer scholarships – but because in a lot of
low-income areas, people just aren't as invested in army service. We'll
occasionally get a kid from a low socioeconomic background who'll
do whatever it takes to enter an elite unit, even if their community
isn't behind them.

Can you give an example?

Alex came from a single-parent home, probably among the lowest tenth
percentile socioeconomically. He worked two jobs to support his mom
and brother and paid for training out of his own pocket. He wanted
to be in Hovlim (Naval Academy), but four months into our course
decided it was too hard because of his work and family obligations.
After a long conversation, I convinced him to continue. Two months
later, he passed the *gibbush* and was offered a spot in the Air Force,
advancing to the final stages of Pilot Training. He ended up serving in
Sayeret Golani (the Golani Special Forces unit), one of the most elite
infantry units in the IDF. And this is a kid who came from nothing.

Have you noticed patterns as to who goes on to pass the *gibbush*?

It's tricky. To pass a *gibbush* you need to stand out, but on the other
hand, you need to be a team player. On a twelve-kilometer stretcher
run, are you going to rush to finish first, or drop back and help others?

The ones who succeed balance individual strength, social awareness, leadership, and collaboration.

What's the biggest weakness you see in your students today?

Mental strength. We get great athletes who've played competitive sports their whole lives, but who quickly learn that physical strength isn't enough. The physical strength you need to pass a *gibbush* is different from what's needed on a soccer field, especially with all the added pressure of being yelled at by commanders who monitor your every move.

One of the most famous expressions in the IDF is *"Hakol b'rosh"* (It's all in your head), the idea that anything is possible if you want it badly enough. What's your take?

It's true, but only if you come with a solid physical base. We each have our limits, and sometimes our bodies' physiology simply can't be overcome.

Last question, Daniel. Some say that today's young Israelis aren't as high-minded and eager to serve the country as previous ones. Do you see that?

[*Laughs.*] Every generation believes the next isn't as selfless or hard-working as they were. At the same time, there's no denying that this generation is a bit more self-centered. This is, after all, the iPhone generation, what we call in Israel *"dor ha'interest"* [the generation of self-interest]. Plenty of my students are eager to serve the country, but a lot are concerned with what's in it for them and how being in a good unit will reflect on them personally.

Do you address this?

Absolutely. I try to teach that self-interest is a recipe for disappointment. Let's say there's a strong kid who, for whatever reason, doesn't pass the *gibbush* and ends up in Artillery, which is respectable but not elite. If he's only in it for himself, his service will be miserable. But if he's motivated to do his part to contribute to the country, his service will be more fulfilling. That's another way our mission has changed. We're not just preparing them to pass the *gibbush*, we're teaching kids about teamwork, perseverance, and how to handle disappointment – concepts needed not only to succeed on a *gibbush*, but in life.

Lone Soldiers: "What the Hell Was I Thinking?"

"Pretty much every day, at least once, I ask myself what the hell I'm doing here," says Nadav, a twenty-year-old Special Forces soldier in the Nahal Brigade. "Freezing my ass off, getting yelled at, doing push-ups in the mud. What was I thinking?"

One of Israeli soldiers' favorite pastimes is wishing they'd never joined the IDF. Even the most *moor'al tzehubim* (see the army slang section below) reminisce about how good life was before.

For most, though, not joining the IDF was never an option. They're drafted after high school and don't have a choice.

But for Nadav and other lone soldiers who volunteer, "What was I thinking?" is a legitimate question. Unlike their Israeli peers, Nadav and other lone soldiers had options: college, internships, backpacking Europe with friends…all of which sound like more fun than guard duty in the rain.

In the 1990s, there were a few hundred foreign lone soldiers at any given time in the IDF. Today there are more than thirty-five hundred from seventy countries, including places you'd expect (Australia, England, New Jersey) and plenty you wouldn't, like Sweden, India, and Denmark.

So why the growing number of volunteers?

"They come for one of three reasons," says Greg, a former lone soldier from Arizona and a volunteer at the Michael Levin Lone Soldier Center in Jerusalem. (Because Greg works for the Israeli Department of Defense, we've changed his name.) Since he helped found the Center in 2009, Greg has advised hundreds of future, current, and past lone soldiers. "Reason one is that they have a very strong connection with Israel. They went to day school and summer camp, visited with their families, and consider army service a crucial piece of their Jewish identity." These soldiers, says Greg, are highly motivated and do well.

Other lone soldiers are fleeing anti-Semitism. "Kids from Turkey, the Ukraine, France, and other places sense that it's no longer safe or easy to be Jewish in their home countries," Greg says. "They're doing what Jews have done for thousands of years, seeking a better life." For them, military service is an obligation, not a choice. Not surprisingly, they often struggle during their service.

And then there are those in a third category, a group Greg delicately calls the lost souls. "These are kids who for whatever reason never found their places back home," he says. "They tried college, but it didn't work. Or they came on Birthright, met soldiers, and saw a chance to be a hero, so they signed up."

It's the soldiers in this last category that Greg is most concerned about. "They believe that serving in the IDF will somehow save them. But of course, it won't."

But let's back up: What is a lone soldier? To qualify as a lone soldier, one must prove that neither parent is living in the country. Once certified, lone soldiers receive benefits including extra pay, a rent subsidy, days off to handle errands, gifts or gift cards on Jewish holidays, and – best of all, many say – a month off to visit their parents in their home countries, for every year they serve.

Despite these perks, lone soldiers face enormous challenges, including language barriers, culture shock, and having to do one's own grocery shopping and laundry on Friday afternoons before the country shuts down for Shabbat.

But the biggest challenge is the one that defines them: being alone. "The IDF is an army of the people," says Greg. "When Israelis are drafted, the expectation is that their family will take care of them on the weekends, attend their ceremonies, and be a support system during their service. Soldiers and their families go through the military together."

Lone soldiers lack this kind of family support. Greg experienced these challenges himself, which is one of the reasons he helped found the Lone Soldier Center with Harriet and Mark Levin, parents of Michael, a lone soldier from Philadelphia who was killed in 2006.

Today, the Levin Lone Soldier Center has branches in Jerusalem, Tel Aviv, and Beer Sheva. Professional and volunteer staff offer current, former, and future lone soldiers everything from Shabbat dinners to guidance on which units to serve in and how to negotiate a medical leave. They assist soldiers after their service, too, helping them find jobs and housing after release.

This last point is particularly important: about half of all lone soldiers return to their home countries after their service, sometimes to pursue higher education, but often because they struggle to acclimate into Israeli society.

These days, lone soldiers get an enormous amount of respect from their sabra commanders and peers, and from Israeli society in general. But it wasn't always so.

Until 2006, most Israelis weren't aware that there was such a thing as lone soldiers. That changed when Michael Levin died in the Second Lebanon War. Fearing there wouldn't be enough people at Michael's burial to say Kaddish (the memorial prayer for the dead), friends of the Levin family put out word that volunteers were needed to make a minyan (a prayer quorum of ten) at the funeral. Social media got wind of the story, as did Israeli news. Thousands of Israelis turned out, packing the cemetery out of respect for a soldier they didn't know but whose story touched them. (See "Trait #5: *Kehilatiyut*" in chapter 1 for Michael's mother's words on this event.) Lone soldiers finally had a face and narrative.

Today, some fear that Israel's love for lone soldiers may have swung too far. Judy Maltz, a journalist with the daily *Haaretz*, investigated a spate of recent suicides among lone soldiers who'd come from overseas. What she and her team discovered was shocking.[2]

"The IDF wasn't vetting these kids ahead of time," Maltz says. "If

2. Judy Maltz and Yaniv Kubovich, "What's Killing Israel's Lone Soldiers?", *Haaretz*, August 25, 2019.

they'd looked, they would have seen that many of them weren't fit, neither mentally or physically, for military service anywhere – much less in an army where they don't speak the language and have no family support."

Maltz received hate mail after the article was published, she says. Not that she was surprised. "The IDF in general, and lone soldiers specifically, in many ways epitomize the Zionist dream," she says. "People don't want to hear that the system might be flawed, or that not all lone soldiers are as heroic as we believed."

In addition to disenchanted readers, Maltz also heard from the IDF. Not to criticize her, but because they were concerned. "They wanted to debrief me and asked for advice. I urged them to vet candidates more closely. 'Who are these kids coming from overseas? What kinds of families are they coming from?'"

"It's hard enough to serve in the army when you grow up here, when you speak Hebrew and have your mom to cook for you and do your laundry when you're home for Shabbat," she adds. "For lone soldiers, the challenge is triple. Not everyone is cut out."

Despite the hardships, lone soldiers tend to look back on their service as a net positive, according to Greg. Even those who leave Israel are proud they served and say that, given the chance, they'd do it again.

"At the end of training, after a year and a half of blood, sweat, tears, and a broken hand, I stood on Mount Kanaim and received my Special Forces pin," recalls the lone soldier Nadav. "I'll never forget that moment. Everything I went through was worth it."

Sabra Lone Soldiers: No Family, Near or Far

It's hard enough to be an IDF soldier when Mom and Dad are back in America or Mexico, thousands of miles away.

But what about when there is no mom or dad at all?

"The soldiers I work with are the anonymous ones," says Liora Rubenstein, senior director of programs at the Levin Lone Soldier

Center in Tel Aviv. "Everyone loves the brave lone soldier who left America to fight for the Jewish homeland. But the sabra whose father is in jail and mother is an alcoholic? No one really wants to hear about that."

Of the seven thousand lone soldiers in the IDF, the majority (about four thousand) are the ones we talked about in the previous chapter – foreign-born volunteers who join the army by choice. These are the ones we typically think about when we hear the phrase "lone soldier."

But there's a whole other category of lone soldiers who don't get nearly the attention the foreigners do – the three thousand sabras who are *chosrei oref* (without home support). They are born and raised in Israel, but, for various reasons, are completely cut off from their families. According to Rubenstein, almost all have at some point suffered serious trauma. She estimates that one-third of the sabra lone soldiers she works with have been sexually abused; others endured years of physical and emotional abuse and neglect.

This, according to Rubenstein, is why almost nobody knows about them – and why it's difficult for her to fundraise. "The stories of the soldiers I work with go against everything we want to believe about Israel," she says. "A complete contradiction to the Zionist dream."

Rubenstein, who works exclusively with *chosrei oref* lone soldiers, exudes the confidence of a battle-worn officer who's seen it all, and with good reason: she herself was a high-ranking officer in the IDF Manpower Division, eventually rising to the rank of lieutenant colonel. According to her, the soldiers who come to her generally fall into two categories.

"About 30 percent are what's known in Hebrew as *noar b'sikun* [endangered youth]," she says. "They come from broken homes in which one or both of the parents were mentally ill, addicted to drugs, or in and out of prison. Sometimes all three."

The other 70 percent, Rubenstein estimates, are from the ultra-Orthodox Haredi community and were kicked out of their homes or, more commonly, chose to run away. "In many cases, they are one of ten or more children in a family that was incapable of taking care of even one," she says. "Some decided they didn't want to be religious or chose to flee an arranged marriage."

And lately, Rubenstein says, she's seeing a new phenomenon among ex-Haredi lone soldiers. "Almost half of the former Haredi kids in our shelter are gay, lesbian, or trans," she says. "These are kids for whom there is no place in the ultra-Orthodox community."

So they leave.

Like all lone soldiers, those who are *chosrei oref* face enormous challenges – the first of which is proving they're a lone soldier at all. "It's much easier to show up at the Ministry of the Interior with a copy of your parents' foreign passports," Rubenstein says. "But how do you prove you're cut off from your family and have no support? How do you prove that your father abused you and you haven't spoken to your mother in two years?"

Certainly, their families have no incentive to back up their children's claims. Parents don't want to face criminal charges of neglect. "When an IDF social worker shows up, the parents say, 'What are you talking about? Of course he still lives here! Come, I'll show you his room.'"

There's also a much deeper challenge these lone soldiers face, one that their foreign volunteers do not: shame. "For foreign volunteers, being a lone soldier is a badge of honor," says Rubenstein. The soldiers she works with, meanwhile, don't want anyone to know they sleep in a shelter on weekends, and six months ago they were living in a car.

Rubenstein tells the story of one soldier who carefully devised a plan for how to keep his story secret. Friday mornings, when his platoon went home for Shabbat, he stood outside the base at the bus stop with a packed duffle bag, with everyone else. "Then, once everyone left, he went back inside the base and slept in an empty food pantry behind the kitchen. He didn't want his commanders and peers to know he had nowhere to go."

Given the challenges they face and their traumatic pasts, why do these *chosrei oref* soldiers join the army at all? Why not just apply for an exemption, or simply dodge the draft?

"Because of my circumstances, I could have easily gotten out of it," says Oria, who left home to escape her abusive father when she was seventeen. "But I believe in service. I wanted to do my part for the country. And find a new path for myself, too."

This, according to Rubenstein, is why many sabra lone soldiers ultimately serve. "When I speak with them, what they almost always tell me is that they see the army as a chance for a fresh start. Like it or not, the army is still the entry card into everyday society.

"And that's why the soldiers I work with are so special," Rubenstein says. "For them, military service is more than just an obligation. It's a chance for a better life."

Army Slang: No Kitbag Questions Allowed

When your commander bursts in at three a.m., flicks on the lights, and shouts, "Wake up, Platoon 2. You have six minutes to be outside in full uniform, flak jackets and helmets on, two canteens filled to the top. Any questions?" and your platoon-mate asks, "Do we need to polish our boots?" he just pissed off your entire unit by asking what's called a *she'elat kitbag*. (See full definition below.)

Think twelve years of Jewish day school taught you to speak Hebrew? Good luck understanding *slang tzva'i* (military slang), the unofficial language of the IDF – the rich collection of idioms, acronyms, and made-up jargon that soldiers use to describe the misery, frustration, and, every so often, euphoria of army life.

Slang tzva'i is a living, breathing language; expressions come and go, and words that were considered *lahit* (hot) thirty years ago are now regarded as *sachi* (nerdy, square, or passé).

Because the majority of Israelis serve in the IDF, many of these expressions and words become part of everyday Israeli vernacular.

Here are twenty-five of the most popular in use today. Learn them, and you'll sound as *moor'al* as a *chapash tza'ir* who just got word he's been granted a *shoosh cornflakes*!

"Ad matai?" "Until when?" True meaning: "How much longer do I have to take this crap?" When combat soldiers reach their final six

months of service, they are allowed to start saying *"Ad matai?"* as an expression of frustration.

ahfter. An outing with the platoon to a restaurant, shopping mall, or other fun place, usually in the late *after*noon or early evening.

asur chalav. "No milk." A game within certain infantry units, in which soldiers in their first two years of service are prohibited from drinking milk or adding it to their tea.

asur sapah. "No couch." A game in which soldiers with under two years of service are prohibited from sitting on sofas in the lounge.

atik. "Ancient." A soldier who's been in the army so long he's allowed to say, *"Ad matai?"*

chafsan. A soldier who always tries to find a way out of hard work and assignments. From the Hebrew verb *l'chapes* (to search).

chamshoosh. An extended Shabbat leave, from Thursday through Sunday. Derived from combining the words *chamishi* ("fifth," denoting Thursday) and *shishi-Shabbat* ("sixth-Sabbath," Friday through Saturday).

chapash. Acronym of *chayal pashut* (simple soldier). One who is not a commander or officer but just serves in the rank and file.

chazir tash. (See also *"tash"* below.) A soldier who does everything he can to secure good conditions such as extra visits home, sick days that he's not really entitled to, and extra sleep for supposed injuries.

dir balak. "Pay attention" (Arabic). Used as a warning: "Be standing in formation at six o'clock, and *dir balak* if even one of you is late!" The connotation is more like, "If you screw this up, may God have mercy on your soul, because I'll make you regret it!"

"Eifo kulam?" "Where is everyone?" True meaning: "Why am I still stuck doing this?" A soldier whose service has been extended, perhaps because of time in jail, is allowed to say, *"Eifo kulam?"* out of frustration, but only after his original discharge date has passed.

flaht. From the English "flat." Used exclusively in the Navy: a soldier who is constantly getting seasick and needs to lie down.

jobnik. A soldier who does a non-combat support job, usually a menial task.

killer. Also just in the Navy. The opposite of *flaht*: a soldier who never gets sick, despite the choppy waves.

laga'at b'kir. "Touch the wall." The exact midpoint of a soldier's service, based on the image of a wind sprint in which you start at a line, sprint to the wall, touch it, and then run back. A soldier who has touched the wall now has less than half of his or her service left.

mischakei pazam. The unofficial games soldiers in combat units play to keep themselves entertained, such as *"asur chalav"* and *"asur sapah."*

moor'al. "Poisoned." A soldier who is incredibly over-the-top gung ho about serving.

oketz. "Sting." A soldier who tries to "sting" the system by getting undeserved goodies, such as doctors' visits, sick leave when he's not really sick, and exemptions from kitchen work and guard duty.

pazam. Acronym of *perek zman mizari* (remaining period of time), denoting how long a soldier has served and level of seniority: a soldier with a lot of *pazam,* called a *pazamnik,* is given more respect in the unit and can boss around soldiers with less. (See *"mischakei pazam"* above.)

ravoosh. A super-extended Shabbat leave, from Wednesday through Sunday. Derived from combining *revi'i* ("fourth," i.e., Wednesday) and *shishi* ("sixth," Friday). Like Bigfoot, the *ravoosh* is often spoken about but rarely seen.

rosh gadol. "Big head." A soldier who goes above and beyond. Ask him to clean the toilets, and he'll also scrub the sink and shower and mop the bathroom floor.

rosh katan. "Small head." A soldier who does the minimum. Ask him to carry a crate of dishes to the dining hall, and he'll carry the crate and leave it outside the kitchen door, not bothering to make sure someone knows he left it or to bring back the emptied crate.

shachor. "Black." The opposite of *tzahov* – a soldier who can't stand the army and can't wait to get out.

shavuz. Like *shachor*, a soldier who is broken, depressed, deflated, and can't bear the thought of another day in the army. Short for *shavur-zayin* (broken male anatomy).

she'elat kitbag. "Kitbag question." When your commander says, "Be outside the bunk at six o'clock sharp!" and a soldier asks, "Do we need to bring our kitbag?" the commander will inevitably say yes – causing the soldier and his platoon-mates to work harder than they would have if he'd simply kept his mouth shut. It's the equivalent of raising your hand and asking the teacher if there's homework. Also known as a *she'elat tza'ir*, a question asked by a young or new soldier.

shoosh. A standard Shabbat/weekend leave, Friday morning through Sunday morning.

shoosh cornflakes. An early-morning *shoosh*, getting you home in time for breakfast.

shoosh Kiddush. A late-starting *shoosh* that gets you home Friday afternoon, just in time for Shabbat Kiddush with your family.

tash. Acronym of *tena'ei sherut* (conditions of service) – the rights to which soldiers are entitled, including sick days, Shabbat leaves, and hours of sleep. (See *"chazir tash"* above, a *tash* pig.)

tzahov. "Yellow." Like *moor'al*, except almost always used to describe a soldier who signs on for more time in the army to become an officer.

tza'ir/tze'irah. "Young one" (male/female). A soldier at the very beginning of service.

tchupar. A treat, usually in the form of an unexpected privilege, such as an extra day at home for leave or a platoon movie night.

wasach. An Arabic word meaning a high level of the cool factor, such as state-of-the-art equipment, high-end gear, or a big, high-tech rifle.

ya sayar. An expression said to a soldier who only looks out for himself. For example, he walks outside the bunk to inconspicuously open a bag of chips that he won't have to share with anyone. Comes from Sayeret, meaning Special Forces; they have a reputation for going into a war zone and, although doing exemplary work, looking out for themselves without regard for any allies in the vicinity.

If you're thinking of joining the IDF, we highly recommend you be a *rosh gadol* and learn these words, now.

If you don't, *dir balak*! But don't say we didn't warn you.

Face to Face with Undercover Operative Shir Peled

Meet Shir Peled, the first female operative in the Mista'aravim, the IDF's undercover anti-terrorism unit made famous in the TV show *Fauda*. From 2002 to 2005, during the Second Intifada, Shir participated in high-stakes undercover combat missions in the West Bank. In these operations, Shir was dressed as a religious Arab woman and, underneath her galabia (traditional female dress), armed.

We talk with Shir about how she came to the unit, why she describes her military experience as a "battle on two fronts," and the one operation she'll never forget.

Joel Chasnoff: First and most obvious question: How does a nice Jewish girl end up in one of the most dangerous combat units of the IDF?

Shir Peled: In a way, I was prepared from the beginning. I majored in theater through high school, and my dream was always to be an actor. At the same time, my father was a commander in the anti-terrorism unit of the police, so I grew up in that world too. But then, in my final year of high school, I was coming home from school and got on a bus in downtown Jerusalem like any other day. At the exact moment I boarded the bus, a terrorist walked behind the vehicle and blew himself up. The people sitting in the back were killed or injured; I was physically unharmed. That was the moment that my worlds collided, and I knew I would pursue this kind of work in the IDF.

You were among the first cohort of women to do this job. Logistically, how did it happen?

I drafted into Magav, the Border Police. At that time, the military's undercover ops were becoming less effective, due to their difficulty of blending into small villages where everyone knew each other. They realized that embedding women into the operations would provide a much greater element of surprise that nobody would expect.

Say more – why, as a woman, would it be so great?

In traditional Arab society, a religiously dressed woman in public is practically invisible. I could walk around undetected, and I didn't even need to learn Arabic, since it's forbidden for a man to speak to an Arab woman in religious dress.

What was your job like, as an undercover operative?

I was not a spy. A spy has one identity that she lives under, uninterrupted, for a long period of time. As a *mista'arevet*, I did short stints in enemy territory. My job was not to interact with the locals, but to be there, on the ground, the moment the operation took place and assist in the arrest, and then get out as quickly as possible.

Can you tell me about your most memorable mission?

One I'll never forget was the turning point in my service. Our mission was to capture a group of known terrorists who, at the time, were in the process of planning a large-scale attack. Our intelligence determined that they would show up at a specific restaurant at a particular time for a meet-up, and that they would be armed. My job was to be on the ground, near the restaurant, and notify our capture team the moment the suspects arrived, at which point our anti-terrorism unit would sweep in, make the arrests, and I would depart. At the moment of the operation, one of the terrorists suspected something. He evaded capture and ran at us. As my partner tackled him, I drew my gun and neutralized the suspect, saving the operation.

In the midst of the chaos, involving several undercover fighters, how could you tell our side from theirs?

That's a great question that I can't answer. Let's just say we have our ways.

In your lectures, you describe your military experience as a "battle on two fronts." Why?

There was, of course, the battle my unit fought against those trying to carry out terror attacks on Israeli citizens. In addition, I had to fight a battle within the military itself, to prove I belonged in a male-dominant system that didn't value women. When I began my training, the men didn't take me seriously. I was with two other women, and we were constantly under the magnifying glass. Many questioned whether a woman would be up to the task, would have the *beitzim* ["eggs," Israeli slang for testicles]. They feared we would be a liability, neither emotionally nor physically strong enough to be effective. They wondered if, in the moment of truth, would I react, on instinct, like a guy, or would I freeze? That's why the mission I described was so important to me. When I saved the operation by drawing my weapon and neutralizing the suspect, it showed my colleagues and myself that not only could I be effective, but even more so than a man because of the unexpected inclusion of a woman.

Let's turn from the IDF to Israeli society in general. What improvements need to be made in the conditions of women?

For me, the biggest factor is *shivyon miktzo'i*, equality in the workforce. Israel is still very much a male-led society, one where women in powerful positions feel they need to behave as men. Look at the female TV news anchors – they all dress and speak in traditionally masculine ways. I think about Knesset member Limor Livnat. The first time she came to the podium to address the Knesset, she came in a dress and long hair, and the men all laughed. She was shaken. To fit in, she changed her speaking voice, attire, and hairstyle. Which brings us back to the army: it's such a central part of Israeli society, the doorway to many other aspects of life. If women aren't treated equally there, we'll have no place as equals in society as a whole. When a seventeen-year-old woman goes to the IDF induction center, she needs to believe that anything is possible, just like for her male counterparts. Only when that is true will she also believe that the sky's also the limit throughout Israeli society. It's the right thing to do and makes for a stronger, better Israel.

LGBTQ IDF

In 1993, the same year the United States barred openly gay and lesbian service members from the military, Israel declared that, effective immediately, "out" soldiers could serve in any branch of the IDF, including elite combat units and intelligence units requiring security clearance.

"That decision was huge, both for gay and lesbian teens and for Israeli society," says Ofer Neuman of Israeli Gay Youth (IGY), where he runs workshops for LGBTQ teenagers. "The message was clear: gay and lesbian kids can serve Israel as much as anyone else."

In numerous studies, the IDF has been named one of the gay-friendliest militaries on the planet. Uniformed soldiers march openly in the country's Pride parades. In 2017, Brigadier General Sharon Afek became the first openly gay member of the IDF General Staff (equivalent to the Joint Chiefs of Staff in the US). And as the *Times of Israel* reports, IDF personnel and their same-sex partners are afforded full spousal rights, including parental leave and deceased service member benefits should they be killed in action (this despite the fact that the State of Israel itself doesn't recognize same-sex marriages performed in the country).[3] In a recent development, the IDF now deems gender-affirming surgery a "necessary medical expense" for transgender soldiers who wish to transition and will even pay for it.[4]

But what's it actually like in the *shetach* ("out in the field," no pun intended) for gay and lesbian soldiers? To what extent does their experience match policy?

Corporal Ori, a twenty-year-old communications instructor in a co-ed combat platoon, came out in eleventh grade. When his draft notice came in the mail, he made a bold choice: to be open about his sexuality from the start. "This is who I am, and if anyone doesn't accept me, that's their problem, not mine."

Ori describes his service so far as *"madhim"* (amazing), though he acknowledges that he can't necessarily assume that all LGBTQ soldiers share his experience. He also notes that the ease of his experience affords him a huge sense of relief. "I didn't know what to expect," he says. "But it's been wonderful. The guys treat me as an equal – no abuse, no jokes in the showers. And the girls love it because they have a guy who loves gossiping about the other guys in the unit as much as they do."

The only hiccup happened when "I saw this cute guy on Grinder and swiped right, but the bastard didn't swipe me back. A month later, we got a new officer – and it was him."

Not all soldiers describe their experiences so glowingly. For Sergeant Eliza, an openly bisexual lone soldier also in a co-ed unit of the Search and Rescue Division, IDF service introduced her to something she'd never encountered: ignorance.

"It started the first night of basic training," she says. "We were sitting in a circle, introducing ourselves. When I said I had a girlfriend, the other girls went nuts."

From that moment, Eliza says, the girls labeled her *halesbit* (the

---

3. Judah Ari Gross, "For Gay Soldiers, IDF Seen as 'More Progressive' than the State," *Times of Israel*, July 5, 2016.
4. Ibid.

lesbian). And while the men in the unit had no issue, the women were, in Eliza's words, "obsessive."

"When I'd walk by, they'd say, 'Here she comes, the *lesbit!*'" she recalls. "Throughout basic training, they asked intrusive questions like 'How do two women do it?' and 'Who's the man in the relationship?' There isn't one, I'd say; that's the point."

Eliza notes that the treatment she received was not malicious or homophobic, but rather, rooted in curiosity. "I was the first 'out' person from the LGBT community most of them had ever met," she says. On the one hand, they watched her every move, but at the same time, Eliza says, they were incredibly loving. "When I spoke to my girlfriend on the phone, they'd always ask to say hi. 'Shalom, *mami!*' they'd say, using the popular term of endearment. 'We're taking good care of your sweetheart.' But in our room, they'd cover up when undressing and freak out if I was naked in front of them. It wasn't spiteful. Just ignorant."

One reason openly gay and lesbian soldiers continue to face challenges is that in many ways, Israel is still very traditional, especially outside of what's known as *"bu'at Tel Aviv"* (the Tel Aviv bubble). Old-fashioned values of what coupledom and marriage should look like are very strong. Compounding this prevalent frame of mind is the fact that in Israel, the concept of political correctness doesn't exist. (In Hebrew, *"homo"* is commonly used as a term for a gay man – neither as a means of empowerment nor as a slur, but because *it's literally the Hebrew word for a gay man.*)

This is very much the polar opposite of the attitude in the northern California environment where Eliza was raised – not that this justifies her experience. "I've been out and secure with myself since age fifteen, so I wasn't hurt," she says. "But I was disappointed. Growing up in a strong Jewish and pro-Israel community, I came here expecting something else. Something more."

As for Ori, his experience couldn't have been more different. In June, Pride Month, Ori asked his commanders for permission to conduct a seminar on the Pride movement in Israel. They agreed, and it was, as Ori describes it, a fantastic night – one of the highlights of his service.

If he has any complaints, it's that he, and not the IDF, had to deliver the seminar. The IDF Education Corps regularly facilitates seminars on everything from mental health to seatbelt safety and drug abuse, delivered by enlisted soldiers, reservists, and outside professionals. The current curriculum does not, however, include education on LGBTQ issues. "There needs to be more awareness," Ori says. "About soldiers like me and LGBTQ in general."

Ofer Neurman at Israel Gay Youth agrees. "We still have work to bring Israel to a place of equality and inclusion," he says. "Every soldier deserves to be in a safe, welcome environment, no matter who they are or where they serve."

In 2000, Bar-Ilan University professor Danny Kaplan and Haifa University professor Amit Rosenmann asked more than five hundred male IDF soldiers if they knew – or thought they knew – of a gay soldier serving in their platoon. They then asked them to rate eleven attributes related to what Kaplan calls "social cohesion" in their unit, such as "enjoy[ing] doing things together" and "admiration."[5]

Kaplan found that in combat platoons where soldiers knew or thought they knew that a fellow soldier was gay, the sense of social cohesion was just as high as in units where they did not know of a gay comrade. "This was, I believe, the first study anywhere to demonstrate that social cohesion does not suffer from the presence of gay soldiers in combat platoons," Kaplan says.

Upon learning of his study, some in the US suggested that America should therefore allow gay and lesbian soldiers to serve openly. Kaplan warned, however, that his findings might be specific to Israel.

"When it comes to gays in the military, the IDF is very tolerant," Kaplan says. "We are, after all, *tzava ha'am* – a people's army. Soldiers

5. Danny Kaplan and Amir Rosenmann, "Unit Social Cohesion in the Israeli Military as a Case Study of 'Don't Ask, Don't Tell,'" *Political Psychology* 33, no. 4 (2012): 1–18.

expect to meet and serve alongside the various cross sections of the population, including soldiers whose sexual orientations are different from their own."

Especially, Sergeant Eliza says, in the Search and Rescue Division – a unit whose nickname among many soldiers, she says, is "Olim v'Homo'im" (Immigrants and Gays).

Face to Face with IDF Special Forces Psychologist Glenn Cohen

> The soldier shall make use of his weaponry and power only for the fulfillment of the mission and solely to the extent required; he will maintain his humanity even in combat. The soldier shall not employ his weaponry and power in order to harm non-combatants or prisoners of war, and shall do all he can to avoid harming their lives, body, honor, and property.
>
> *– Tohar haneshek (the code of "purity of arms"), an excerpt from The Spirit of the IDF, the Israeli army's doctrine of ethics*

Glenn Cohen is a former Israeli Air Force pilot; hostage negotiator; special forces operative; and, in his most recent position, chief psychologist of the Mossad, the country's infamous national intelligence agency. In 2011, Cohen was involved in the efforts to release Gilad Shalit, the IDF soldier kidnapped by Hamas and held for five years in Gaza. Ultimately, in exchange for Shalit's return, Israel released more than one thousand Palestinian and Arab-Israeli prisoners, almost all convicted terrorists, who were altogether responsible for the murder of more than five hundred Israelis. We spoke with Cohen about the Shalit negotiations, the evolution of the IDF's code of ethics known as "*tohar haneshek*," and the moral quandaries the army faces daily.

Joel Chasnoff: Ethicists will be discussing the Gilad Shalit prisoner exchange for years to come. What is the calculus of saving one actual life versus risking some unknown number of lives that one

day *could* be lost by releasing convicted terrorists, some of whom have already gone on to kill again?

Glenn Cohen: This is a question that the IDF and the Jewish people have grappled with for years. The Rambam [Rabbi Moses Maimonides, born 1138] wrote that of all the mitzvot [commandments], the most important is *pidyon shvuyim*, bringing back captured soldiers [literally, "the redemption of captives"]. But it goes back even further. In the book of Genesis, four kings capture Lot, Abraham's nephew. Despite his small army, Abraham was determined to wage war on the kings in order to save his relative. In a daring operation, they rescued Lot and saved his life. This value remains a fundamental part of the Israeli ethos, to the degree that when a soldier is captured, the entire country feels the trauma. When Gilad Shalit sat in captivity for five years, he was everybody's son.

And yet some people would argue that this policy only reinforces terrorism, sending the message that terrorism works.

Let's be clear: *"Tohar haneshek"* is both the IDF's strength and weakness, and our enemies know it. Their primary goal is to kidnap Israeli soldiers because they know we'll do anything to bring them home. But I would argue that as opposed to making us weaker, it's the key to our success. Our soldiers go to battle knowing that we will always have their backs, that they will never be abandoned.

add new question and answer:

On October 7, 2023, Hamas kidnapped more than two hundred Israeli citizens and took them hostage in Gaza, with over a hundred still being held as of this writing. Does this change the Israeli public's attitude toward hostage negotiation and the price we're willing to pay to bring them home? Should it?

The savage attack and kidnapping has led Israelis to adopt what psychologists call a "split" approach – in this case, feeling that one is either in favor of a deal to release the hostages ASAP, or in favor of continued fighting in Gaza with the goal of dismantling Hamas.

Psychologically, it's better and healthier to have what's known as an "integrative" approach to the negotiations: let's make a deal to release the hostages, while at the same time enabling us to continue to dismantle Hamas. Difficult as it may be, the Israeli public needs to hold both of these ideas in our collective mind.

You mentioned "*tohar haneshek*," which is the guiding principle behind the Shalit decision and IDF conduct in general. How is "purity of arms" different from other countries' military codes of ethics?

Other armies' codes exist for worst-case scenarios, instructing soldiers how to behave in captivity to avoid giving up secrets, and how to treat enemy POWs. In the IDF, "*tohar haneshek*" is more than just a code of conduct, it's a code of ethics that dictates how to behave day in and day out. It's founded on the principle of the sanctity of life, both the enemy's and our own.

How does that actually play out?

For our own soldiers, it means two things. First, that a soldier should not risk their own life, or the lives of those under their command, unnecessarily – we don't send our soldiers on suicide missions from which they have no chance to return. Second, they know that should they be captured by the enemy, the State of Israel will do everything in its power to bring them home. In terms of our enemies, "purity of arms" means doing everything possible to avoid civilian casualties, even if it puts our own soldiers at risk.

Can you give an example of how that might play out today?

Countless times, our soldiers have encountered suspicious women or teenagers who appeared to be wearing something bulky under their coats, most likely a suicide bomb. Instead of "shooting first, asking questions later," our soldiers held fire and suffered injury when, indeed, the suspects blew themselves up. More recent examples come from our operations against Hamas in Gaza. It would be easy to just level a building to "send a message." Instead, before bombing, we confirm that a target absolutely poses a threat. We then drop pamphlets and notify our intelligence contacts on the other side and implement a policy called "knocking on the roof," in which we release non-explosive devices on the roofs of buildings about to be bombed, so that civilians have a chance to flee ahead of time. Of course, enemy combatants will flee too, but that's a price we're willing to pay.

So was "*tohar haneshek*" a principle from the start?

Most definitely not! In the early years of Israel, it was important to demonstrate that our young country was strong and resilient. Getting captured by the enemy was not an option; in those days, the IDF ethos was "The last bullet in the cartridge is saved for me." Any soldier who did get captured would face court-martial if they ever managed to return home, which they wouldn't, because they most likely would have been tortured to death.

In your lectures, you talk about how "*tohar haneshek*" isn't confined to the military, but permeates Israeli society as a whole. How so?

Every Israeli citizen knows that wherever they are in the world, Israel has their back. In 1976, when Israelis were held hostage in Uganda, we didn't wait for the international community to intervene, we flew to Entebbe and rescued them ourselves. If you're hiking in the Himalayas today and there's an avalanche, you'll know we're sending a helicopter to save you. It's the Israeli way.

Last question, Glenn. Some of the incredible moral quandaries you've described sound like no-win situations, a choice between multiple bad options. What would you say to Israel's critics?

Like any army, we are made up of living, breathing human beings, complex individuals with free will. We are not perfect, nor is the army they together make up. That said, we hold everyone accountable to the ethical code we believe in, and we continue to navigate unwaveringly by our moral compass.

CHAPTER 7

The Arts, Culture, Sports, and Leisure

ON OCTOBER 7, 2017, ACTRESS GAL GADOT HOSTED *SATURDAY Night Live* and joked during her comedic monologue that the show's writers clearly knew nothing about Israel, because they had her eating hummus in every sketch. Gadot said this on American television, to millions of people...in Hebrew!

Incredible as this was (local media even broadcast the show live so Israelis could tune in), Gadot's appearance on one of the biggest shows in television history was emblematic of a bigger story. Just as Israel has made major contributions to the world in areas such as cybersecurity and biotech, recent years have witnessed the country take steps forward in sports and entertainment as well, be it Olympic athletes accepting their gold medals to the soundtrack of "Hatikvah" or shows like *Fauda* and *Shtisel* becoming international sensations.

In chapter 7, we'll look at some of Israel's success stories in the world of arts, culture, and sport, from the man whose trip abroad led to his changing Israeli music forever to the stand-up comics whose dark humor helps them and their audiences overcome adversity. We'll explore how the country's unique set of economic and political circumstances have forced its artists and athletes to adapt and ultimately thrive.

Finally, we'll examine Israelis' unique relationship with the land itself, from the young hikers who soul-search along the north-south Israel National Trail to the guides who bring the land alive.

Having a homeland of our own has afforded our people not just a sense of safety in body, but also in spirit. Instead of figuring out how to survive, our psychic energies can be fueled into something more transcendent: how to create.

Face to Face with Olympian Yael Arad

On July 30, 1992, twenty-five-year-old Israeli judoka Yael Arad made history by winning Israel's first-ever Olympic medal, at the Barcelona Summer Games.

Today, Arad is president of the Israel Olympic Committee, and she has big plans – both for the future of Israeli athletics and for Israeli society as a whole.

We sat down with Arad to talk about "barefoot sports," what Israel can learn from the British, and how Israeli athletes are like no other.

Joel Chasnoff: It's an honor to talk with you, Yael. That summer of '92, I had just graduated high school and was working at a Jewish summer camp in the US. I still remember the entire camp celebrating upon learning you'd won a medal.

Yael Arad: Thank you! I always love hearing that.

As president of the Olympic Committee of Israel, what's your outlook on the future of Israeli athletics?

First, let's acknowledge that there's a lot to be proud of *now*. We're a young country who have already won thirteen medals, and "Hatikvah" has been played three times at the Olympic Games. That's a big accomplishment.

Looking ahead, I believe Israel can medal in sports beyond the ones we're competing in now. There's a joke in the Israeli Olympic world that we're only good at *sport yachef* (barefoot sports), like judo, sailing, and gymnastics. I want us to medal in sports that require shoes!

Specifically?

Track and field, for example. I also believe we can compete in soccer by developing athletes from our Arab population. Getting them into the system is not only good for performance, but just as important, it's good for Israeli society.

How do we make that happen? What challenges do you and the Olympic Committee face?

It's a matter of resources, which itself is a product of a few things. First, we're a small country. The US has 330 million people, Germany eighty million, and we've got nine million – a much smaller pool to draw from. That's why it's important to bring Arab-Israeli athletes into the fold.

It's also a matter of finances. Because our budget is small, we tend to invest in the sports where we're already successful. We need to choose a few new sports and invest heavily in them. Our model should be the UK. When London was chosen to host the 2012 Summer Games, the British felt an urgency to win and invested heavily in a few sports, like cycling. It paid off – they won several cycling medals at those Games. For us, this means choosing a few sports that we can compete in and investing in them on the youth level. If we create a large pool of athletes to choose from, and then develop them from a young age, we can compete globally.

The other obstacle we face is lack of tradition. I think of Austria, with its great tradition of Olympic skiers, where kids grow up dreaming of carrying it on. We have that a bit in judo, which I'm obviously thrilled to see, but we don't yet have that tradition in other arenas.

What's the response from the government so far?

It hasn't been easy. Budgets are always tight in Israel, and Olympic sports haven't been a high priority. The good news is, the government is realizing that, just like high tech, Olympic athletics are a great way to spread a message of Israeli excellence.

After you won the silver in Barcelona, you dedicated your medal to the Israeli athletes murdered at the 1972 Munich Games. Tell me about that.

Before leaving for Barcelona, I visited the families of the fallen athletes. I sat with them in their living rooms and looked at their photo albums. Twenty years had passed since the tragedy, but for them, the wound was raw. Before I left, the families gave me a small book of *tehilim* [psalms] to take with me. I made a silent promise to myself that if I should be so blessed to win a medal, I would dedicate it to them.

How did it feel when you were actually able to do that?

Incredibly powerful. Anytime a country medals for the first time, the press conference is especially big. When I spoke, the first thing I said was that today we closed a circle in Israeli athletics. Most people didn't know much about Munich, and many had moved on, but I brought attention to it, connecting the worst day of Israeli athletics to one of the best.

Last question, Yael. As a competitor and now president of the Olympic Committee of Israel, you've met athletes and coaches from around the globe. What is unique about our athletes, compared to other countries'?

When it comes to training, we work as hard as any athletes in the world, anywhere. The difference, I think, is the sense of national pride. We're a young country, a small country, one where the connection between personal and national achievement is very high. An accomplishment by one Israeli is an accomplishment by all. You see this in high tech – when an Israeli company goes public or sells for a billion dollars, it's a source of pride for everyone. I'm sure American athletes appreciate that they're competing for America, but I don't think there's the same degree of personal connection, that sense that every individual American is counting on them and personally invested in their performance. But that's part of who we are. We're a country where civilians give years of their life to the army or National Service for something bigger than themselves, and that idea has filtered into athletics, too.

Jews around the world feel it, also. Our athletes visit Jewish communities around the globe. I did this myself when I was twenty-five, and it was amazing. Like you and your friends at summer camp, so many people told me that they were rooting for me. It was like getting a big family hug.

That's what's unique about the Israeli athlete – that sense of who we're competing for. Munich is always in the back of our minds. When we compete, we're competing for them, also. Not just as Israelis, but as Jews.

Israeli Comedy: Never Too Soon

"My old army unit recently invited me to perform," says Israeli comic Eldad Shetrit in a recent show. "It went great. Afterwards, the officer in charge asked for my current address so he could send a thank-you note. A couple days later, I got a letter summoning me to a month of reserve duty."

From the Eastern European shtetl to twenty-first-century Hollywood, Jewish comedy has always centered around otherness. Lenny Bruce comparing "Jewish and goyish," Mel Brooks singing "The Inquisition" in *History of the World: Part I*, Woody Allen buying white bread and mayonnaise to become more genuinely Gentile in *Hannah and Her Sisters*...For Jews, being funny relied on playing a part in someone else's story.

"This all changed with the birth of Israel, redefining Jewish identity and comedy along with it," according to comedy writer and director Alon Gur Arye. "For generations, Yiddish speakers used humor to cope with anti-Semitism, persecution, assimilation, and living as outsiders in societies that, at best, tolerated us, and at worst, tried to kill us." Upon gaining independence, the paradigm of Jewish humor suddenly changed: If being funny used to be a response to persecution, outsider status, and powerlessness, what happens when Jews become the majority, and "the other" is replaced by "us"?

One popular subject of this new iteration of Jewish humor was the nascent state itself. Ephraim Kishon's 1964 film *Salach Shabati* starred Chaim Topol as a Mizrachi immigrant trying to acclimate after making aliyah with his family. (Topol would later play Tevye in the 1971 film *Fiddler on the Roof.*) Kishon satirized kibbutz inhabitants for their indifference toward their poor Mizrachi neighbors; the Jewish National Fund for their practice of selling trees to Diaspora Jews (in the movie, their representative dupes wealthy American philanthropists into believing they're getting their own personal forests); and Israel's primitive voting process, in which citizens stick notes in envelopes (though simple-minded Salach bungles it after politicians promise him the world).

Salach Shabati won the Golden Globe for Best Foreign Film, notable as an example of mainstream Jewish humor that, for once, was *not* about a world out to get us. It's also considered the first of what Israelis call a *Bourekas* film, the series of movies named by director Boaz Davidson after the Israeli *bourekas* pastry. Just as Italian movie makers had directed "Spaghetti Westerns," Davidson nonchalantly compared his own films to one of our local foods, and the name stuck. According to Gur Arye, in Israel's early stages, the Ashkenazi class dominated politics as well as popular culture, with Mizrachi and Sepharadi immigrants banished to the background. *Bourekas* films like *Charlie v'Chetzi* and *Chagigah b'Snooker,* meanwhile, brought the overlooked Mizrachi and Sepharadi cultures into Israeli society.

In these films, the protagonists were more than willing to make fun of themselves. "The character Salach Shabati – played by an Ashkenazi actor, by the way – not only can't read, but doesn't realize that politicians are trying to buy his vote, because he doesn't even know what a bribe is!" Arye says. But more importantly, like countless protagonists before them, the dark-skinned, underpowered Mizrachi characters took revenge by punching up, only this time toward fellow Jews in the form of the snobby Ashkenazi elite. Like so much of Jewish comedy, the humor was a classic setup of David versus Goliath – except in this case, Goliath was also Jewish, and his family ate gefilte fish.

Keeping the jokes "within the family" is fun and provides psychological relief, but these new Jews didn't suddenly rid themselves of two-thousand-year-old baggage. This is best seen in a famous sketch from the 1990s show *Hachamishiah Hakamerit* (*The Kameri Five*), whose writers included well-known author (and child of Holocaust survivors) Etgar Keret. In the sketch, Germany hosts a track and field competition that includes a delegation from Israel. Moments before the race begins, two Israelis accost the German referee to negotiate a head start for the Israeli athlete, attempting to guilt him into conceding by referencing the Holocaust: "Haven't the Jewish people suffered enough?"

Gur Arye explains, "This sketch illustrates the paradoxical, split identity of today's Israelis: on one hand, a world leader in security, innovation, and technology, backed by a powerful military; on the other, a people whose deep scars and trauma remain, just eighty years removed from unspeakable tragedies." This subject appears again in one of Keret's most beloved short stories, "Shoes," a seriocomic tale about a little boy whose parents bring him a pair of Adidas sneakers from Germany, the very country where the boy's grandfather perished in a concentration camp.

The best Israeli comedy also highlights one of Israel's continued dichotomies: the clash between our history as victims and our current place of power. One of the great modern sketch shows, *Hayehudim Ba'im* (*The Jews Are Coming*), "documented" Israel's singular use of the death penalty in the Eichmann Trial and how the guards, having no prior experience, required fifty-three efforts to get it right.

Like much of Israel, the comedy scene is influenced by globalization. Thanks to Netflix (and local performances of top acts such as Jerry Seinfeld, Sarah Silverman, Chris Rock, and Bill Burr), interest in comedy has exploded, with nightly stand-up shows now occurring in Tel Aviv as well as Jerusalem, Haifa, and Be'ersheva.

Just as the sarcasm and dryness of British humor reflect the personalities of the joke tellers, Israeli stand-up is truly Israeli: blunt, in your face, and no topic out of bounds. "This entire country is living in a state of post-trauma," says comedian Sigal Kahana. "War, terrorist attacks, the Holocaust. Dark humor is about more than just comedy. By laughing at pain, we heal."

That said, there is one subject that even Israeli comedians won't touch: Israeli soldiers killed in combat.

"It's just too personal," Kahana says. "Every family in Israel knows at least one soldier who fell. As a comedian, I'm willing to go almost anywhere. But not there."

Kaveret: The Revolution in Israeli Music

On June 7, 1973, music journalist Benny Dudkevitch arrived at Tel Aviv's Beit Hamoreh event hall, skeptical of the idea of rock music in Hebrew. Hours later? "I couldn't believe my own ears. For the first time, Hebrew rock actually sounded legit!"

Like the hottest TV shows and consumer goods of the 1960s, the massive social and cultural changes sweeping the world did make it to Israel, just with a bit of a delay. Though the Beatles weren't allowed to visit during their heyday (the government agency that approved foreign artists' visits feared they would influence Israeli youth for the worse), the musical revolution they helped launch did make it, thanks to the genius of Danny Sanderson, an Israeli guitarist who discovered their music while abroad. To this day, Kaveret, Israel's answer to the Beatles as well as the group who would win over skeptics like Dudkevitch, remains the most popular band in the country's history.

"The story goes like this," Boaz Cohen, radio DJ for 99FM, begins...

In the 1960s, young Danny moved with his family from the Tel Aviv suburbs to New York City for his father's work assignment. According to Cohen, while attending the city's High School of Music and Art, Danny was exposed to the experimental sounds that defined the era – groups like the Allman Brothers, the Beach Boys, and the Beatles. Their albums sounded nothing like Israeli music, which, at the time, still resembled pre-1948 Russian military songs: lyrics infused with patriotism, messages relating to wartime and peoplehood, often performed by army troupes, and showcasing the instrument that speaks to rebellious teens across generations (the accordion!).

Inspired by these groups and the sounds of Jimi Hendrix and Cream, Danny picked up guitar and piano before returning to Israel in 1968 for his mandatory military service. He joined one of the musical troupes, the Nachal Band, performing around the country for soldiers and continuing to experiment with electric guitar. Cohen says it was Jimi who inspired Danny to arrange and play the Hendrix-esque guitar intro to what would become one of Israel's biggest hits, "Shir l'Shalom" (A Song for Peace); the song became even more meaningful when on November 4, 1995, Prime Minister Yitzhak Rabin led a chorus of thousands of Israelis in unison, just minutes before he was assassinated.

In the Nachal Band, Danny befriended and performed with several musicians who would become his future bandmates: singer Gidi Gov, bassist Alon Olearchik, guitarist Efraim Shamir, and drummer Meir Fenigstein. In addition to their music, their special chemistry and sense of humor drew attention from Galei Tzahal, the army radio station, which invited them to record humorous skits for their listeners. Their recurring segment was called "Pinot Poogy" (Poogy's Corner), named for the skits' main character, an alter ego of Fenigstein.

After their discharge from the army, the quintet, along with guitarist Yitzhak Klepter and keyboardist Yoni Rechter, joined forces under the name Kaveret, the Hebrew word for "beehive," due to the band's collaborative nature (as any apiculturist worth her honey knows, bees work together when building a hive). In 1972, Kaveret recorded their first album based on a rock opera called *Sippurei Poogy* (Poogy Tales) that Sanderson wrote. Though they ditched the opera format at the suggestion of producer Avraham Deshe, they retained the name for the album. (Deshe clearly knew what he was doing; he was also the founder of Israel's most popular comedy troupe, Hagashash Hachiver!)

Sippurei Poogy was released in November 1973, just weeks after the

Yom Kippur War. Forget the O_2 Arena or Madison Square Garden; Kaveret's first album tour took place at military posts as part of their reserve duty. Sanderson told the *Forward*, "Our songs were a form of escapism for people who were very concerned and frightened because of what was going on. When reserve soldiers finally got to return to their civilian lives, they remembered those magical moments with us..."[1] Just like the movie *Givat Halfon*, Kaveret and their upbeat and humorous spirit gave the country a reason to smile after the heartbreak of war.

Not only did *Sippurei Poogy* sound musically different, its lyrics were nothing like traditional army music, instead focusing on everyday life, such as the fun classic "Shir Hamakolet" (The Grocery Song), about a man who recalls the woman he fell for at the supermarket.

Perhaps their most popular hit, "Yo Ya" tells of a man sentenced to the death penalty, sitting in the electric chair and wishing he could switch seats because *"Meshaneh makom, meshaneh mazal,"* or "if you change your place, you change your luck." Incredible guitar, an awesome drum solo, and a Talmudic quote that's still used today, all in one song. No wonder Jewish kids around the world still dance to it in their youth movement or summer camp.

In *Sippurei Poogy*'s first year, Israelis bought seventy thousand copies, a massive number for a population of just three million people, and another seventy thousand by the end of the 1980s. Kaveret became the country's top-selling band of all time, as well as the Israel Broadcasting Authority and Galei Tzahal's "Band of the Year" for four straight years. They even represented Israel in the international Eurovision competition (but lost to ABBA – no shame in that!).

In the coming years, Kaveret released two more albums before splitting up in 1976. By then, their place in history was sealed: their progressive music with a local twist inspired generations, with parents passing their albums down to their children. The band reunited at least

---

1. Tal Kra-Oz, "Is It Last Dance for Kaveret, the 'Israeli ABBA'?," *The Forward*, June 13, 2013.

five times between the 1970s and 2010s, including a concert that drew half a million spectators, the largest show ever in Israel.

With what we know about the "typical rock and roll lifestyle," it may be fortuitous that all seven band members are still alive into their seventies. But Kaveret weren't typical rock stars. If anything, they were typical Israelis.

"Their vibe was 'We're not that important and don't take ourselves so seriously,'" says Boaz Cohen. "They were all family men – you could say 'nice Jewish boys.'"

Who *wouldn't* want to hang out with them a bit at the local grocery store?

Givat Halfon Einah Onah: Cult Film Classic

In February 1983, over a hundred million Americans watched the series finale of M*A*S*H, the largest broadcast television audience in history aside from Super Bowls. Part of what made the show about army life so popular was its ability to find humor in a topic normally considered stressful, traumatic, or grim. No less important was its mocking of an institution that wasn't often poked fun at: the military.

Perhaps then it's not surprising that in survey after survey, Israelis have chosen the military comedy *Givat Halfon Einah Onah* (Halfon Hill Doesn't Answer) as the most popular movie in the country's history.

Released in 1976 by filmmaker Assi Dayan, *Givat Halfon* isn't known for its sophisticated plot. Despite its mostly negative reviews and miniscule budget, the film airs on TV every Yom Ha'atzmaut for viewers who can recite every ridiculous line, whether or not they were alive when it came out.

To understand the place of this beloved movie in Israel's history, you need to understand some backstory. From 1967 to 1973, Israelis existed in a state of euphoria as a result of their speedy, unexpected, and absolute victory in the Six-Day War. For the first time in its young history, Israel felt invincible. Much of the credit for this stunning triumph was attributed to the military cunning of the country's defense minister, Moshe Dayan, remembered today as much for his iconic eye patch as for his bravery and leadership against overwhelming odds.

Cut to October 1973. Still riding the wave of euphoria, Israel was caught off guard when neighboring armies attacked on Yom Kippur, the holiest day on the Jewish calendar. More than twenty-five hundred lives were lost from a population of just three million people. Prime Minister Golda Meir resigned and many blamed Defense Minister Dayan for failing to foresee the impending attack and the massive casualties.

Which, oddly enough, brings us back to *Givat Halfon*, directed by Assi Dayan – none other than the son of the decorated legend. As film director Alon Gur Arye explains, young Assi couldn't have been more different from his dad; his rebellious nature even led to him spending a month in military prison after fleeing during an exercise. A pacifist who wanted nothing to do with a military career, Assi instead pursued acting and film directing, making lowbrow anti-establishment films. After the Yom Kippur War, Assi Dayan reported for *miluim* and spent months in the Sinai Desert without any action, except for what would be his greatest idea: a movie about a bumbling reserve unit, spending months in the Sinai without any action.

The movie's full title, *Givat Halfon Einah Onah*, is a send-up of the classic 1950s army movie *Givat 24 Einah Onah* (Hill 24 Doesn't Answer). The cast featured the three members of Israel's most popular comedy troupe, Hagashash Hachiver (The Pale Tracker), mentioned in the previous chapter. The troupe's sketches about war, the Hebrew language, the absorption of immigrants, and other pressing current events were, and still are, iconic to Israeli pop culture. With a budget of only $300,000, the film shot in three weeks. The highlight of the movie is the interactions among the three "Gashashim" and between them and the authority figures, most of whom present the IDF as something less than the mythical army it was thought to be.

The anti-army ribbing starts with the opening theme song, presenting reserve duty not as a well-oiled military exercise, but as an opportunity, when life gets too stressful, to "immediately take a vacation and go to *miluim*." How a vacation? In the days before cell phones, internet, and constant connectivity, leaving home for the Sinai became a reprieve from the stresses of your normal routine. Or in other words, as the song continues, if someone needed to reach you from "the bank or tax authority, Halfon Hill isn't answering."[2]

In one of the most famous scenes, a visiting colonel tests a soldier's readiness for an enemy attack (spoiler: the soldier fails miserably). When the colonel asks what to do if the Egyptians approach the outpost, the soldier answers, "what we did in '56." When the colonel demands more details, the soldier references the war of '48. Which was? His bluff called, the soldier finally admits defeat with the classic line, "Thirty years, who can remember?"[3]

As the movie nears its climax, the bumbling soldier referenced above accidentally crosses the Egyptian border while attempting to go fishing in the desert – a scene that, as it turns out, was inspired by reality. A year and a half before Dayan died in 2014, he created an autobiographical documentary called *Hachayim k'Shmuah* (*Life as a Rumor*) in which he told the story of how, when he was just three years old and Israel not even a year old, young Assi walked out of his family's Jerusalem residence and accidentally crossed the armistice line into Jordan. When the Jordanians found him and figured out the identity of the boy's father, they contacted Israel immediately and returned him as soon as they could.

Good satire makes its audience think and reflect. Just as *Salach Shabati* was the first movie to poke holes in the mythology of the Zionist project (see the "Israeli Comedy" section above), *Givat Halfon* did the same for the IDF. To make fun of the most lauded institution was previously unthinkable, but then again, so was military defeat – a prospect Israel almost faced in the Yom Kippur War. In the new period

---

2. *Givat Halfon Einah Onah*, directed by Assi Dayan (1976), translated by the authors.
3. Ibid.

of national self-critique that war inspired, Assi Dayan stepped up, eager to change the way his fellow Israelis viewed the army. Essentially, he sought to push them to reexamine their collective heart and soul.

As Gur Arye explains, "There was almost no precedent for this kind of self-reflection before 1973. Think about how long ago this was; we didn't even elect our first right-wing government until after the movie came out!" Still, *Givat Halfon* wasn't considered a political movie as much as a funny one.

Explosive Stand-Up: Mohamed Namaa

Camel Comedy Club, Tel Aviv. A Thursday night.

The comedian takes the stage. He glances at his watch, nervous. Then he leans into the microphone and whispers, "This'll have to be quick. In five minutes, I need to get back to my village. There's a curfew."

The audience chuckles – uncomfortably at first, then more freely as they realize it's *okay* to be laughing at this.

"So, a bit about me," he says. "I'm thirty-two years old and single, but *me'od mevukash* [a Hebrew expression that can mean either 'in demand' or 'wanted']. Not by women," he explains. "Just *wanted*. You've probably seen my face on a poster."

If you're struggling to find the humor in these jokes, it helps to know the comedian in question is Mohamed Namaa, an Arab Israeli from the village of Deir al Assad in the Galilee. Since 2009, Mohamed has performed all over Israel, cracking jokes in both Hebrew and Arabic about checkpoints, mistaken identity, and other pitfalls of living as an Arab in a Jewish state.

Like Chris Rock, Lenny Bruce, and other comics who've challenged the status quo, Mohamed uses his ability to make people laugh to explore dicey topics that are harder to talk about offstage – things like racism, religion, and ethnicity. In one very short joke, he confronts one of the biggest open secrets in Israeli society today: in a country whose founding ethos was the pioneer spirit, almost all menial labor is now performed by immigrants and the underprivileged.

"Any Ethiopians in the crowd?" he often asks during a show. Then, after silence that often follows: "Shit. *Again* I have to clean?"

Mohamed didn't always want to be a comic. Until he landed a job as a cook at the Camel, he didn't know what stand-up was. "That first night on the job, I felt like a boy in Disneyland," he says. "I didn't care who would be in the crowd or what language I'd have to perform in, I just wanted to try."

He started in Hebrew, not only because the Arabic comedy scene is all but nonexistent, but also because it's the language he's most proficient in (which isn't uncommon for Arabs immersed in Israel's Hebrew-centric culture). As for material, he did what all great comics do: he talked about what he knew. In his case, the hardship of being Mohamed in Tel Aviv.

"The other night, I was at a restaurant with a Jewish friend," he jokes a few nights later at the Comedy Bar. "At the end of the meal, the waitress asks if she can bring me anything else. I say, '*Lo, ani od shniyah mitpotzetz!*' [No, I'm about to explode!]. I got tackled by three waiters and spent the night in jail. I won't be eating *there* again."

These days, Mohamed is an audience favorite. His humor is original and smart, and he's got that mysterious quality showbiz folk call "likability." Crowds also love that he's familiar with Jewish tradition. In one recent show, he raised his water bottle and recited the Jewish *shehakol* blessing before taking a drink, cracking up the audience at the sheer absurdity of it.

Much of Mohamed's knowledge of Judaism is a result of the two years he spent living and working with Jews during his Sherut Leumi (National Service, an alternative to the military obligation that has recently welcomed a growing number of Arab Israelis). From ages eighteen to twenty, Mohamed was a translator for Jewish doctors and Palestinian patients at the Tel HaShomer Military Hospital in Ramat Gan.

He lived in an army-issue apartment with an Ethiopian immigrant, an ultra-religious Jew from Bnei Brak, and a gay Jewish man from Tel Aviv. No, they didn't all "walk into a bar," but they did struggle to cohabit, each being from a community the others hadn't previously encountered. After the initial domestic friction, they put their differences aside and lived in peace. Mohamed credits this experience with teaching him that coexistence is not only desirable, but possible.

That's not to say he hasn't had his share of tough moments onstage. One night, a Jewish audience member stood up and, to collective shock, yelled at him to "go home to your village." Instead of kicking him out, Mohamed asked the man where he was from and why he felt that way. Over the course of their exchange, it came out that Mohamed and the heckler were both fans of the Beitar Yerushalayim soccer team, whose rowdy supporters unabashedly chant *"Mavet la'Aravim!"* (Death to Arabs!) Mohamed said that he, too, goes to the games, but yells *"Mavet lanu!"* (Death to us!). The improvised joke broke the tension and bridged the gap. A few minutes later, the heckler stood up and shook Mohamed's hand.

Mohamed's comedy plays largely on his conviction that Muslims and Jews have more in common than they think, so there's no reason the two peoples can't get along. "All over the country are Arab doctors, pharmacists, soldiers," he says. "I've been alive thirty-two years, and not for one second did I want to hurt someone because he's a Jew."

He sees comedy as a vehicle for change, especially when it comes to calling attention to what Arabs in Israel experience every day. Before closing out his set, Mohamed likes to ask the audience if he can take a selfie with them – so he has an alibi if the Mishmar Hagevul (the IDF's notorious Border Patrol Unit) comes asking.

The line always gets a big laugh. As he snaps the picture and leaves the stage, audiences typically rise to their feet and applaud. Their standing ovation is more than just an acknowledgment of a great performance. It's recognition that there's another narrative happening in the country, one that millions of people live with but most Israelis rarely think about. With his stories and jokes, Mohamed Namaa gives audiences a glimpse into this alternate world.

But only for a few minutes. He needs to get back to his village.

The Basketball Miracle: Aulcie Perry

Lake Placid, New York, February 1980. With ten minutes to go in the final period, team captain Mike Eruzione flicks a wrist shot past Russian goalie Vladimir Myshkin to give the underdog US Olympic Men's Hockey team a 4–3 victory against the powerhouse Soviets. It's one the most iconic moments in American sports history, known today as the "Miracle on Ice."

Three years earlier, hundreds of thousands of Israelis flooded the streets after an upset victory of their own against a different Russian squad, when underdog Maccabi Tel Aviv defeated the mighty CSKA Moscow at the 1977 EuroLeague Basketball semifinals. After the game, the team's star, Tal Brody, raised the trophy and exclaimed, in one of the most memorable quotes in Israeli history: "We are on the map and we are *staying* on the map, not just in sports but in everything!"

"The Miracle on Hardwood," as it was later dubbed in the documentary *On the Map*, gave Israelis a much-needed lift after the horrific murder of eleven Israeli athletes at the 1972 Munich Olympics and forever changed what Israelis believed was possible in the context of sport.

How did a squad representing a country of four million overcome the longest of odds to defeat a global superpower? Teamwork. Motivation. And outstanding defense from a heretofore unknown player named Aulcie Perry.

"I was never *not* tall," Perry says over drinks at a Tel Aviv café. A native of Newark, New Jersey, Perry was six foot five by the age of thirteen. But more important than his ability to dominate the boards was his unwavering drive to prepare. "We were joking around in practice one day in high school when my coach pulled me aside and said something I would never forget," Perry recalls. "'Failing to prepare is preparing to fail.' From that moment on, I committed to preparing for every game the best I could. Before and after school, in the rain and snow. Every single day."

America's eighth-leading scorer in college, Perry was signed to a pro contract by the Virginia Squires of the American Basketball Association – the same team that kick-started the careers of Dr. J and George "the Iceman" Gervin. A year later, Perry joined the NBA's New York Knicks after competing against Phil Jackson, Walt Frazier, and Bill Bradley for a roster spot. Before he could play a single game, however, he was cut and his spot given to yet another future Hall of Famer, Spencer Haywood.

"At that point, I assumed it was over," Perry says. But fate had other plans.

One afternoon, while playing in Harlem's Rucker Park, he was spotted and offered a job by Shamluk Machrowski, general manager of Maccabi Tel Aviv. "The team could have been in Siberia and I would have gone," Perry says – so badly did he want to play. Machrowski made clear that Perry's services would be needed for only six games, since Maccabi had never escaped the first round of the EuroLeague tournament and surely wouldn't this year, either.

Alongside proven scorers like Jim Boatwright and Lou Silver, plus star Tal Brody, who'd been drafted into the NBA but had instead chosen to make aliyah, Perry was happy to embrace a role that other players might have shied away from: "dirty work," as Perry phrases it. Tough defense, rebounding, whatever it took to win.

In 1977, few expected Maccabi to advance deep into the EuroLeague tournament, much less make the semifinals. When they did, they faced a formidable opponent: the Russian CSKA squad, known as "the Red Army" for its connection to the Soviet military. CSKA contained no fewer than six players from the 1972 Olympics gold medal squad (the one that beat the US, thanks to multiple controversial late-game calls). And like that matchup and the "Miracle on Ice," this contest would be more than just a game. At the height of the Cold War, Israel and Russia had no diplomatic relations, and Russia was actively selling weapons to Israel's enemies. Because they refused to come to Israel or grant visas to the Israeli team, the game was played in a tiny gym in Virton, Belgium.

Though new to Israel, Perry didn't lack motivation. "It bothered me the way Russia treated its Jews," he says. "But what really got me was how no one in Israel expected us to win. People were begging us to please only lose by twenty, not fifty." Plus, Perry couldn't forget what happened to America's team in 1972.

As Perry recalls, coach Ralph Klein's pre-game speech didn't focus on Xs and Os. Instead, he talked about how Russia had treated his ancestors and how his father was murdered in Auschwitz. "He told us if we won the game for the people of Israel," Perry says, "the country would never forget it."

In a strange irony, the day of the game – February 17, 1977 – Prime Minister Yitzhak Rabin had planned to resign over a scandal related to an undeclared US bank account. But because the entire country was tuned in to the game, Rabin had to delay his announcement.

Instead of a political resignation, the Israeli public was treated to a masterpiece, led by Perry. He blocked several shots under the basket, forcing his opponents to stay outside the paint. He hit a few key jumpers of his own, drawing out the Russian defense and opening up the court for Boatwright. In this David-Goliath matchup, it was the former who once again triumphed, 91–79. Tal Brody said his famous words on Israeli television, and the entire nation emptied into the streets in mass celebration.

After the team beat Italy's Pallacanestro Varese in the final, Perry began getting calls from NBA teams. To the surprise of many, he turned them down. "I couldn't risk getting cut again and losing another season. And I was happy in Israel!"

So happy, in fact, that he converted to Judaism and officially made aliyah after the 1978 season. He's lived here ever since, today coaching Maccabi's youth team and working in the PR department for the professional club.

As we wrap up our interview and ask for the check, the older barista asks the younger one, "Do you know who that is?"

When the younger shakes his head, the older says, "Google 'Aulcie Perry.' Or just 'Israel basketball miracle.'"

Face to Face with Culinary Expert Inbal Baum

Meet Inbal Baum, founder of Delicious Israel, a culinary tourism company specializing in food tours and cooking workshops for those who want to encounter Israel not just through their ears and eyes, but with their taste buds. The daughter of Israeli parents, Baum grew up in Washington, DC, but came to Israel every year with her family. In 2009, burned out from working long hours as a Holocaust art reparations lawyer in New York City, she decided to fulfill her lifelong dream of making aliyah.

We talk to Baum about what makes food "Israeli," the link between climate and produce, and why, despite its simplicity, she'll never get tired of hummus.

Benji Lovitt: It used to be that "Israeli food" meant falafel, but that's no longer the case. What *is* Israeli food, and what's changed over the years?

Inbal Baum: So much! You could write an entire book on the evolution of "Israeli food" from the time of the Bible to the Ottomans and Arab influences until today. One of the biggest factors is immigrants: over the last century, Israel has absorbed people from all over the world, each bringing their own rich mosaic of spices, traditions, and techniques. We're also a very innovative society – so when you put our incredible local produce and ingredients into the hands of young chefs who aren't afraid to experiment, you end up with exciting and creative combinations.

Speaking of which, one of Israel's most famous chefs, Eyal Shani, has become known for his cauliflower recipe. What makes something that sounds so simple so special?

For starters, he doesn't overdo it. Eyal allows the ingredients to speak for themselves. He takes a head of cauliflower, farmed the same day, boils it, massages it with olive oil, roasts it in the oven, adds a dash of salt, some *techinah* [tahini], and done. But there's another reason: the world has changed. It used to be that people wanted long, formal four-hour meals with exquisite technique, but nowadays we crave fresh, local, and accessible food. And with the increased value placed on health, locality, and sustainability, especially by vegetarians and vegans, his recipes really hit the spot.

When people visit Israel for the first time, what about the food scene surprises them?

Just as tourists are often shocked to discover the diversity of the people, the diversity of food is also unexpected. On a single street in Jaffa, you find Arab cuisine next to a hipster coffee bar, Turkish *bourekas*, craft beers, Greek souvlaki, and of course, fresh fish direct from the historic ancient port of Jaffa just a few blocks away. You also have fourth-generation mom-and-pop shops literally next to Michelin star restaurants. Which is why, to really define Israeli food, we have to take it apart piece by piece and put it back together into something that really doesn't exist anywhere else in the world.

Can you give an example?

Take a Japanese restaurant in Israel. An Israeli chef might add the Japanese citrus yuzu to a ceviche dish, using local Israeli ingredients to add a Japanese twist, creating something vibrant and new. A chef in a classic Japanese restaurant wouldn't be willing to take the same risks, bound by the tradition of their food. Israeli chefs aren't afraid to take those traditions and get a little wild with it. This is what I alluded to when I talked about the culture of innovation: "Israeli food" no longer only describes food that's been eaten in the Middle East for thousands of years. It's becoming a label to describe the local flair or stamp that our restaurants are adding to their favorite international dishes through spices, local ingredients, and experimentation.

At the same time, many people outside of Israel still think of Jewish food as bagels and lox – but you can barely find bagels in Israel. Why is that?

First-time visitors might come here wanting lox and bagels for breakfast, but then what do they see at the hotel breakfast bar? *Shakshuka.* When you compare the flavors, spices, and even the colors of Ashkenazi food with Sepharadi/Mizrachi dishes, the latter wins. It's also an issue of climate. Ashkenazi foods came from Eastern Europe and Poland, where the temperatures were colder than the Middle East, where Mizrachi food originates. You can definitely find some great Hungarian blintzes here, but that's not what you're going to want on a daily basis in hot, humid Israeli weather. Of course, Sepharadi and Mizrachi cultures have *chamin*, their version of cholent [a heavy, slow-cooked Sabbath stew], but it's going to have more spice and flavor with cinnamon.

Earlier, you mentioned "local." What's unique about Israel's local ingredients?

We have to start with the produce. When you taste a tomato in Israel, it tastes like a tomato! That's not a given around the world. Because Israel is so small, fruits and vegetables travel no more than six hours to get to the market. In other places, produce may travel much further or even be shipped from overseas, which means it may not yet be ripe when picked and no longer fresh upon arrival. And then of course there's the soil.

Soil? What's special about that?

Fifty percent of the land here is desert, less than ideal for farming. But because of local tech companies and commitment to research and investment, we're able to make the land more arable through creative solutions like fertilizer compounds. The weather is a factor, too. In Israel we have what's known as microclimates – this means that on any given day, you can find cold weather in the north and extreme desert heat in the south, which allows us to grow a wide range of crops. And, of course, if you're going to talk about "local," it's also about our diverse spices. Think about all the countries from which *olim* came, along with their favorite spices and blends – places like Iran with Persian lemon, Iraqi *baharat*, Yemenite *hawaij*, Turkey and its unique combinations of spices for shawarma spice, and Russian dill.

I notice you didn't mention America.

The "Everything Bagel" spice at Trader Joe's is pretty good.

Bagels again!

You can take the girl out of America…

Israel, and Tel Aviv specifically, is one of the most vegan-friendly places around. Why are there so many vegans in Israel?

More and more people around the world are becoming vegans for health and environmental reasons. While many would consider that a major lifestyle change, it's much easier here. The Mediterranean diet has never revolved around meat; people in this region have always favored what the local land most efficiently produced, which is fruits and vegetables. That makes becoming vegan much easier. And that breeds habit – we're not eating bacon for breakfast, we're eating salad three meals a day. And when you have high-quality fruits and vegetables grown nearby, it's easier. It also helps that hummus is a superfood – it's got *techinah*, chickpeas, your protein, your carbs, and it's filling, so you don't feel like you have to have meat instead.

There are so many best-selling Israeli cookbooks today all over the world. How did this happen?

Yotam Ottolenghi and Sami Tamimi in London, Michael Solomonov in Philadelphia, Adeena Sussman in Israel… Their books really captured the Israeli soul. They also hit at a time when the "food porn" culture of Instagram and social media took off, bringing together people who are excited about exploring new foods. And it's not only Jews using these books. All types of people are discovering the Mediterranean diet, especially as the world becomes more health conscious. And of course, it's the variety – an Israeli cookbook can have Turkish, Syrian, and Iraqi recipes, all in one place. In the past, you might have had to buy three separate books.

Last question: What's your favorite food these days and why?

I know it's so cliché, but hummus. The Middle East is the only place where they serve hummus warm, straight from the pot.

And if I asked you to recommend one hummus place, what would you say?

Oh, please don't! There are so many wonderful out-of-the way hummus spots, each of them different. But if I had to choose, it would be Abu Hassan in Jaffa. It is truly soul-filling!

And a superfood!

All the better.

Tour Guides: Israel Comes Alive

In the early 1960s, Uri Dvir of the Israeli Ministry of Tourism launched the country's first official tour guide course – and not a moment too soon. Tourism would boom following 1967's Six-Day War and the IDF's liberation of the Old City. If it's now cliché to claim that Israel is "where old meets new," these early guides were the ones who facilitated the meeting. Suddenly, thousands of locals and visitors could touch the ancient stones of the Western Wall as their guides connected King Herod's role in the Temple's construction to the soldiers who liberated its remains two thousand years later.

These days, hundreds of thousands of visitors come to Israel on youth trips, pilgrimages, and guided tours. And while their itineraries vary, the one thing they share is that magical encounter with the one who links past and present, ancient and modern: the Israeli tour guide.

"Anyone can recite facts and dates," says Michael Bauer, one of Israel's top guides (he guided rock icon Jon Bon Jovi during one of his visits). "But a good guide turns a place into an experience. That's what I'm after with every one of my groups. I want them to *feel* a place – and have it come alive."

If you've been to Israel yourself, you know how impactful a good guide can be; for many visitors, the guide is not just a part of the trip, but a highlight. From the moment they leap onto your bus outside Ben Gurion Airport, grab the microphone, and shout "Shalom, and welcome home!" the Israeli tour guide is larger than life: charismatic, funny, brilliant, entertaining, and seemingly an expert on *every* square inch of the country. "What's the name of that plant? Which biblical character walked this same trail? What animal made that poop you just stepped in?"

But the tour guide's vast body of knowledge isn't limited to rock formations and historical sites. A good guide also conveniently knows the location of every rest stop in the country. Although why wait in line at the Ein Bokek Aroma Café when your guide can brew Turkish coffee on the burner he just took from his backpack? And are you dreaming, or did she just casually grab something off a nearby tree and *eat it*?! She did – and between bites explains that it's a leaf from a salty caper plant (*tzalaf* in Hebrew), one that appears in the Talmud and was often mixed with vinegar as a remedy for heart problems. Pretty impressive for what guide Hillary Menkowitz calls "nature's potato chips."

Menkowitz, who moved to Israel from Philadelphia in 2010, was herself influenced by the guides she worked with when volunteering with Taglit-Birthright. "My inspiration to become a guide came from seeing their impact on the participants." She shares a story of witnessing the reunion between American participants born in the Former Soviet Union and their Israel-based relatives whom they hadn't seen in decades. "So much rests on the guides to provide the critical context. By explaining the history of Soviet Jewry and helping the entire group appreciate and process these special moments, they transform a powerful individual moment into a meaningful communal experience."

Knowledge and experience aside, what really makes a special guide is their love for the land that is nothing less than infectious. Rabbi Ari Vernon, a rabbi in Houston, Texas, recalls: "In the Ramon Crater, our guide Abe Silver pulled four Crembo treats out of his backpack to demonstrate how the *machtesh* [crater] was formed. By pushing his fingers into the Crembo, he simulated the rain eroding the sandstone and volcanic rock. It's been thirty years since Abe mashed those Crembos together, and I still remember it clear as day."

Like the history of the country itself, the field of guiding has evolved dramatically over the last century. In the late 1800s, according to Bauer, pre-state Israel suddenly became a popular destination to live and visit. Those first guides were fixers – locals who could help visitors navigate, bargain, and handle logistics in the local language.

As the infrastructure of the Jewish state grew, this began to change, thanks in large part to Shmaryahu Guttman, an *oleh* from Scotland who devoted his life to Zionist activism and archaeology. Guttman realized the importance of hiking and exploring as a method of education – the foundation of modern tour guiding – through its principle of "leadership through field education" (i.e., learning the land through your feet).

These days, the official certification course lasts about two years. To become licensed, students must pass intense written and oral exams in geography, religion (Jewish, Christian, and Islamic), history, and current events, as well as demonstrate skill in storytelling and pedagogy.

But the real learning, according to Bauer, comes on the job. Like a comedian, Bauer has tweaked, edited, and tested his material thousands of times. He also relies on what he calls his "toolbox" – a dedicated backpack containing whatever he needs for that day's trek. "I actually have fifteen toolboxes," he says. "One for Jerusalem, one for the Judaean Desert, another one for Druze villages up north. A typical backpack has pictures, poems, a quote from the Bible, a *gaziah* [travel coffee burner]…" As Bauer explains, it's one thing to hear the story about Neve Tzedek, the first neighborhood of Tel Aviv; it's another to hear the story while looking at century-old photographs of that very neighborhood as you walk down Chelouche Street, named for its founder Aharon Chelouche.

This is why the great guides gain near-mythical status among their groups: they're not just information givers, but masters at dealing with psychology and peoplehood. "When I do my job right, my participants will return home with a new perspective on identity," Bauer says. "And they'll leave with more questions than answers."

Expert guides are also able to employ a variety of creative techniques to keep visitors engaged at the end of a long day (or in the wee hours of an early start). Menkowitz shares one of her favorite "gimmicks" for teen groups in the Negev. "We know from the Bedouins that, should you ever find yourself in the desert without water, ibex poop contains enough water and minerals to survive for at least a day," she tells her bus, moments before eating a sample straight from the ground. Little do they know, as they gasp and shriek, ibex droppings look exactly like Klik, the delicious Israeli chocolate candy which her staff surreptitiously planted just minutes earlier.

Before you embark on your own des(s)ert hike, please know that Menkowitz's "survival tip" isn't true, but, as she describes, it's "shtick to keep them entertained within the larger context of appreciating the desert, like how we can live there and use it to its fullest." Nevertheless, just as Abe's Crembo demonstration made an impression on Rabbi Vernon that resonates still today, you can be sure that Hillary's group will never look at Klik (or ibex poo) the same way again.

And where did Hillary learn this trick? From one of her own guides, decades earlier, of course!

The Israel National Trail

In Yonatan Netanyahu's posthumous collection of letters, *Self-Portrait of a Hero*, Operation Entebbe's leader and lone casualty writes of his deep love for the land of Israel:

> I have seen and felt the beauty of the Judean Desert, the might of straight, steep cliffs rising vertically for hundreds of feet with only one thin white trail winding through them like a tiny trickle of water, the beauty of dry, parched earth...[4]

In one of his most moving passages, Yoni describes his desire not only to see the land, but to experience it: "I had to know the country," he writes, "and by 'know,' I mean be familiar with every tree and rock."[5]

---

4. Jonathan Netanyahu, with notes and an afterword by Benjamin and Iddo Netanyahu, *Self-Portrait of a Hero: The Letters of Jonathan Netanyahu (1963–1976)* (New York: Random House, 1980).
5. Ibid.

This very desire has been part of the Israeli ethos since its founding. Shvil Yisrael, the six-hundred-mile Israel National Trail that runs from Kibbutz Dan in the north to the southernmost tip of the country in Eilat, is, for many Israelis, the ultimate way to get to know their country…and themselves.

"I grew up in the Merkaz and didn't truly appreciate or even know the entire country," says Petach Tikvah native Ido Shalom, explaining his motivations for hiking the trail in early 2022. "But more than that, I had some very deep questions about my life. My hope was that over the course of the journey, I would resolve them and learn more about myself."

Though many hike the trail in a group, Ido set off on his own. Like most, he started in the north, where the terrain is easier and the trail closer to small towns where one can replenish supplies. His first big challenge came about two weeks in, when he reached Mount Arbel. "I'm afraid of heights," Ido explains. "There's a point where the trail is so narrow that you literally have to hold on to the mountain in order to pass safely. One wrong step and you're finished."

Only now can Ido laugh. "It's better I didn't know it was coming," he says. "Or I never would have begun."

Hani Ing, a Tel Avivian who now lives in Haifa, hiked the trail not once, but twice. "It was the same trail both times, but the experiences were completely different," she says. "Not because the mountains or rocks had changed, but because I had."

Asked to explain, Hani ponders. "The first time, I was in a group, worried about falling behind and upsetting everyone. The second time, I was older and went alone. I was much more willing to forgive myself for not being perfect. I gave myself days off without judging myself as lazy. I was much more comfortable in my own skin."

Hani adds that on the first hike, she focused on how much of the trail remained. "The second time, at the end of each day, I took stock of how far I'd come.

"It's such a better way to view the trail," she says. "And life."

In 2012, *National Geographic* included Shvil Yisrael on its list of twenty epic hikes around the world. As it happens, the trail was inspired by another entry on the list – the Appalachian Trail in the eastern United States.

In the late 1970s, Israeli journalist Avraham Tamir was hiking the Appalachian when it occurred to him that Israel needed its own signature trail. He pitched the idea to his friend Ori Dvir, director of the Society for Protection of Nature in Israel. The SPNI created some of the trail from scratch and connected pieces of the more than fifteen thousand miles of hiking trails already in place.

It took fifteen years, but in 1995, Shvil Yisrael was born. Israelis' search for connection with the land – and themselves – began anew.

Though Israelis and foreigners of all ages attempt parts or the entirety of the trail, most hikers are young Israelis who are *ben l'ven* (between this and that) – just out of the army but not yet enrolled in university, or just out of high school with their military draft still months away.

In the trail's early days, hikers had to carry enough food and water for the entire trip or, more commonly, stash provisions under rocks and hope they would still be there when they returned weeks later.

Today, technology provides an easier solution: a public WhatsApp group that connects hikers with residents who leave water and snacks in wooden crates near the trail, especially helpful in the desert. When hikers post a message that they've reached the Negev, residents send detailed instructions to the supplies' location. Rarely do the two parties meet; instead, hikers wire a small amount of money through a payment app (about ten shekels per liter of water, according to Ido Shalom) on the honor system.

Another recent phenomenon is Malachei Hashvil (Trail Angels), a community of people who live near the trail and open their homes to hikers who need meals, a place to sleep, and something they don't realize they desperately need: conversation. "These kids are on the trail alone, thinking about their lives. When they come to us, it all comes pouring out," says Adina, a Trail Angel since 2007 who lives just over a mile from the trail in Zichron Yaakov.

Adina estimates that her family has hosted about a hundred people over the years, of whom she's stayed in touch with many. "I've gone to their engagement parties, their weddings," she says. "It's a very special connection."

Yardena, who lives with her husband and two children also in the north, has welcomed travelers since 2012. "Hosting is part of our DNA," she says. "And I think there's something very Israeli about opening your home to others. The sense of adventure, diving in without knowing exactly what's going to happen."

Hosting Israel Trail hikers has been a positive influence on her kids, too. Her children play with the hikers, ask questions, and help out with hosting duties. Hikers enjoy the interaction and human contact.

Yardena has implemented one change over the years. When she began hosting, she realized something that should have been obvious: most hikers stink. Many hadn't showered in days.

Now, when they arrive, Yardena asks hikers to leave their boots on the porch. She then leads them to the guest room and points out the towels on the bed are for them, and, *if they want*, the shower is down the hall.

"I'm gentle about it," Yardena says. "But they get the hint."

Hiking the Holy Land

"I've hiked everywhere," American *olah* Susannah Schild tells me as she ducks under a tree branch. "Ireland, Italy, the Scottish Highlands. All beautiful. But there's just something so wonderfully unique about Israel."

Unlike other interviews for this book, this one came with a packing list: sturdy boots, sunscreen, wide-brimmed hat. "And water, at least a liter," Schild texted just as I left home.

Schild made aliyah in 2003 in her early twenties with her husband and two kids (now six, plus a grandchild). In 2018, on a whim, she started a blog called *Hiking the Holy Land* to share her favorite trails with fellow would-be explorers. Today, Schild's site is the go-to for tourists and locals alike who want to get off the beaten path (and their screens) and connect with Israel like our ancestors did: on foot.

Hiking along the "red trail" next to Nachal Katlav in the Judean Mountains' Sorek Nature Reserve, Schild and I plop down under an olive tree and drink. I tell her that I've driven through this area countless times but never knew these trails were here. And that it's really cool how it feels so secluded here in the shade of this olive tree and yet just a small turn of the head reveals a clear view of Jerusalem's downtown.

"Exactly!" Schild exclaims. "This tiny country has mountains, desert, canyons, streams, all within an hour or two from home. Like hidden treasures," she says, and smiles. "You just have to know where to look."

When I ask how hiking Israel differs from other places, she mentions the variety of terrain and that the country can be hiked year-round. "But more than that, it's the history," she says. "To hike Israel is to step back in time. Literally."

She tells of a recent hike of Itri, a Second Temple-era village near Adulam, forty-five kilometers southwest of Jerusalem. "The area is reminiscent of times past. You see shepherds and ruins, a wine press and a synagogue," she says. "But a closer look reveals caves and tunnels from the Bar Kochba revolt against the Romans, two thousand years ago. Sitting in those caves, crawling the tunnels, you can't help but feel like you're part of their story."

Here are six of Schild's favorite off-the-beaten-path hikes that you won't necessarily find in a guidebook. For driving directions and starting points, search in a map application or, easier, on the *Hiking the Holy Land* website.

Nachal Mishmar (Dead Sea area). "Nachal Mishmar is unquestionably one of my favorite desert hikes," Schild says. "The trail leads through a riverbed canyon that's completely quiet except for the sounds of birds, singing and swooping overhead. The alabaster stone creates a natural climbing trail that athletic types will love. Wear good hiking shoes and bring plenty of water."

Nachal Tavor (Lower Galilee). "In February and March, Nachal Tavor is unequivocally the best hike in Israel," says Schild. "The 6.5-kilometer trail brings hikers into a world of lush beauty reminiscent of Scotland or Switzerland. Aside from green rolling hills, Nachal Tavor features a snaking river, basalt waterfalls, and an explosion of colorful wildflowers during the winter months. Purple lupines abound, along with red anemones, yellow mustard flowers, and pink cyclamen. I recommend sturdy hiking boots and hiking poles. Be advised there are several stream crossings, and you'll need to maneuver carefully to stay dry."

To make this hike child-friendly, Schild recommends driving to the end point and then continuing downhill in your car until you reach the stream, then parking and hiking out to the waterfall.

Nachal Kziv (Upper Galilee). "One of the best summer hikes is Nachal Kziv, a long, shady stream with hanging vines and pools of cool water. To do the full hike, you'll need to leave a car at each end or catch a bus back to your starting point. If you make it a shorter, circular hike, you'll get to visit Montfort, a beautiful ancient fortress that towers above rolling green hills. Either way, bring shoes suitable for water and swim gear." Schild adds that those who wish to stay dry can hike next to the stream rather than through it. "But why would you want to do that?"

Nachal Prat (Northern Judean Desert). "When people ask me for an amazing water hike in Israel, I almost always recommend Nachal Prat," says Schild. "This spectacular desert oasis is easily accessible, just thirty minutes from Jerusalem. There are so many ways to hike at the Ein Prat Nature Reserve, with multiple trails suitable for all hikers. In the summer, fig trees and wild mint adorn the flowing streams and pools. In winter and spring, colorful wildflowers appear."

As Schild notes, Nachal Prat also just happens to be the historical stomping grounds of Jeremiah the prophet and the monk Haritoun – so you'll be walking in famous footsteps! She advises checking the park map to determine which trail suits your skill and fitness level.

Burgin and Itri Trail at Adulam (Beit Shemesh, Jerusalem Hills).
"Beit Shemesh is more than a bustling immigrant city; it's surrounded
by beautiful hiking trails, all under an hour's drive from Jerusalem or
Tel Aviv. One of my favorites in the Adulam Nature Reserve is the loop
between Burgin and the Itri Ruins, two ancient towns that date back
to the Second Temple period. In winter and spring, Adulam comes
alive with lush greenery and flowers. In summer, hike at sunset, when
yellow grasses take on a golden glow." In addition to plenty of water,
Schild recommends bringing a headlamp so you can explore the Bar
Kochba tunnels that lie beneath each of the ancient towns.

Loop Trail at Mount Eitan (Sataf, Jerusalem Hills). "This was one
of the first hikes I fell in love with," Shields says. "The trail at Mount
Eitan loops around the summit of a forested mountain. Along the
trail, you can gaze over mountains in some places and over Jerusalem
in others." Because it's mostly flat, Schild says, you can bring a jogging
stroller for little ones who can't yet (or refuse to) walk. All you need
on this eight-kilometer trail is water and a good pair of sneakers.

What does Schild do on weekends, when she's not leading groups,
discovering new trails, or documenting her latest explorations?

Schild smiles. "Every Shabbat, I walk Derech Avot [the Path of the
Forefathers]. It runs about a kilometer from my house. It's the path
that Abraham walked.

"And like him, I always hike it alone."

Face to Face with *Fauda* Cocreator Avi Issacharoff

After serving in an elite unit of the IDF and working as a journalist
in Arab affairs, Avi Issacharoff used his life experience for an entirely
different goal: to entertain.

In 2015, Issacharoff and writer-director Lior Raz cocreated the
international television hit *Fauda*. Seven years later, they sold their production company, Faraway Road Productions, for 50 million dollars.

Despite this windfall, we took *him* for coffee and were happy to do so, just for the chance to talk with Issacharoff about his curious career path, the differences between the Israeli showbiz scene and Hollywood, and why *Fauda* is at once uniquely Israeli and popular worldwide.

Benji Lovitt: One of Israel's recent successes has been TV shows becoming international hits or being adapted into English. Why do you think *Fauda* has become so popular around the world?

Avi Issacharoff: I think *Fauda* brought a level of authenticity that you don't see in Hollywood. In the first season, we see the hero Doron doing an exercise called *yavashim* (dry training), in which he practices quickly cocking his gun before he shoots, until his hands bleed. This is something never before seen in a Hollywood thriller. In an American movie, the hero would practice at a firing range and, of course, hit the bull's-eye every time.

I won't spoil it, but I remember being surprised by something terrible happening in season 1. It wouldn't turn out that way in an American show.

We didn't make choices to try to get ratings or please the crowd. We wanted to create an adventure that seemed like it *could* be real. One that hits you hard, like reality.

Aside from our reality being different, what Israeli characteristics helped make these shows hits?

I know it sounds cliché, but one of the most formative experiences here is the army. It takes you to places that the average American or European wouldn't go. Dan Senor, the author of *Start-Up Nation*, once said to me, "In the army, every operation ends with a *tachkir* [research or evaluation]." He's right – everyone involved in the operation meets and reviews every tiny thing that went wrong. You go one by one and answer, "What did I do wrong? What did I miss?" in order to improve. Dan realized that Israelis brought this process from the army to the high-tech sector. Instead of patting yourself on the back, you constantly look for areas where you can do better.

There's also something about our storytelling. In Israel, eighteen- and nineteen-year-olds experience things that they don't in the US. I think we tell our stories in a unique way. Who knows, maybe it comes from our Jewish heritage, sitting around the dinner table on Friday night and retelling the history of our ancestors.

I imagine that financial limitations must also have an effect.

Absolutely. Our smaller budgets teach us to improvise, something that is very Israeli. We learned a cliché in basic training that "the battle is the kingdom of the unknown." There's a plan for battle, and there's everything unexpected that happens en route to the target. This happens in TV production all the time. "The car didn't come, the food didn't arrive, the actor is sick, and now you're over budget. Now what to do?" You react, maneuver, and figure out how to get it done.

I think this way actually works better. I worked on a Hollywood show with a budget *twenty-four times* the size of *Fauda*'s. If *Fauda* had had that kind of budget, I'm not sure it would have been any better. The most important thing is the story.

It must be quite an adjustment going from Israel to a Hollywood studio. What other differences were there?

As you know, we communicate very differently. I thought I was a very diplomatic person, fluent in multiple languages, but it was often hard to know what someone in Los Angeles was trying to say. If I'm not on board with an Israeli, I'll just say, *"Mah zeh hashtuyot ha'eileh? Zeh lo oved!"* (What is this nonsense? This isn't working!). In LA, a big director told me, "I'm a huge fan of *Fauda*, I want to work with you!" Then I found out he was trying to blow me off. This is why I feel more at home in Ramallah than in Hollywood. If an Arab says he wants to kill me, I know exactly what he means. He won't say, "You're great! Let's work together!"

Speaking of Arabs, how did they see the show?

There isn't just one kind of Arab or Palestinian, just as all Israelis aren't the same. Some Palestinians call for boycotts and say that the show is supporting the Israeli occupation. Some said we were generalizing about Palestinians. First of all, the show isn't about Muslims, but rather a very specific people in Hamas and Islamic Jihad. Secondly, we did the opposite of generalizing. We depicted them as complex human beings, with wives and kids they love and care for, as well as showing our Palestinian allies in a very positive light. This is something that you didn't see until *Fauda*. I imagine this is why so many Palestinians love the show.

How did you even get into show business? After the army, you first became a journalist. How did that happen?

Like a good Israeli, I didn't plan it. I got two degrees in Middle Eastern studies and almost by accident fell into journalism. Then, by accident, I became a screenwriter.

Not just any journalist, though – a journalist reporting on Arab affairs. What was that like?

The biggest challenge was developing sources on the enemy side, not easy when they assume you've been to the army. So now you have to convince them to talk to you, rather than to a Palestinian journalist. This is where chemistry makes all the difference. I went from hunting them down in Special Forces to interviewing them and drinking coffee with them in their own homes as they became my very good sources. And you actually get close to and feel sympathy for some of them. When my ex-wife was in bad shape in the hospital, a Hamas sheikh would call daily, telling me he's praying for her. He calls for millions of *shahids* [martyrs] to descend upon Jerusalem, and he's praying for *me*?!

How do you explain that?

That even the devil is human. That's what we did in *Fauda*, we humanized the devil. And when you get to know people on the other side, you realize that most of us want the same thing: to make a living and enjoy life. Many Israelis think that all Palestinians want to kill Jews, and they think the same thing about us.

But what about Hamas and fundamentalists?

They're fundamentalists. They want to eliminate the State of Israel. I don't think their officials want to just pay the bills. But most people in Gaza are normal people who think we're the devil for bombing their houses.

Last question, Avi. *Fauda* debuted in 2015, and from what I understand, it's as popular as ever. How do you explain that?

I must say, it is quite phenomenal. There's a TV station in London broadcasting *Fauda* in Farsi to Iranians around the world. Two days ago, I was in the UAE as a guest of one of the Emirati crown princes. He's crazy about *Fauda*. There's even a new adaptation of *Fauda* about the Indian Army fighting Kashmir. How can you explain this? As long as the story remains authentic, I think people will continue to love it.

CHAPTER 8

Making Aliyah

IN 1970, THE COMEDY SHOW *LOOL* AIRED WHAT WOULD BECOME
its most famous sketch. Musical legend Arik Einstein and his comedy
partner Uri Zohar (who would later become an ultra-Orthodox rabbi)
played two Russian immigrants who arrive in Israel by boat, kissing the
shores in euphoria as they disembark. In the next scene, a pair of Polish
immigrants arrive in the same manner, but this time, the Russians, considering themselves well-adjusted *olim vatikim* (veteran immigrants),
look on, mocking them for eating "gefilte fish with sugar," among other
customs. The cycle continues with Yemenites, Germans, Moroccans,
and so on, each pair of immigrants evolving from ecstatic innocence
to territorial condescension toward those who come next.

Fifty years later, this sketch is a classic, both for displaying the
rich ethnic mosaic of Israel's people (and each group's respective
stereotype) and also for tapping into the development of immigrants'
ever-evolving relationship with Israel and their fellow citizens.

In this section, we look at aliyah – the process of immigrating to
Israel – and, just as important, the people who choose to do it. Who
are they? What pitfalls do they encounter along the way? And most of
all, what were they thinking? (Believe us, sabras are constantly asking
new immigrants, "What were you *THINKING*?!")

To be sure, challenges including language, finances, and distance
from family cause a fair number of *olim* to leave.

Still, the majority stay. Those who do often experience a magical moment when they realize that, lo and behold, something inside them has changed: they are now Israeli.

Aliyah: Jump!

Thanks to Israel's Law of Return, any Jew around the world can immigrate to Israel and obtain citizenship pretty much instantaneously. The Knesset passed the law in 1950 to populate its new state and provide the refuge that could have saved millions during the Holocaust.

To make the law as inclusive as possible, the Knesset defined a Jew as anyone who has at least one Jewish grandparent or who converted to Judaism. (According to *halachah*, or Jewish law, meanwhile, a person is only Jewish if born from a Jewish mother or after undergoing conversion. This leads to the thorny situations in which people could legally immigrate, serve in the IDF, and put their lives on the line for the country, but not be allowed to marry here – but that's a different story…)

Since 1948, more than three million *olim* have changed the face of the country not only through their own contributions, but through those of their descendants. Today, droves of French and Turkish Jews are coming out of fear of living an openly Jewish life in their home countries. South African and Brazilian Jews see their respective countries collapsing and want to leave before it's too late.

Still, many *olim* come because of what former Knesset member and Jewish Agency chair Natan Sharansky calls an "aliyah of choice" – a search for a more meaningful life in Israel than in the country they once called home.

Which begs the question: Why? What drives people to uproot themselves and often their entire families, pulling kids out of school, leaving secure jobs and abandoning comfortable and stable – albeit imperfect – lives?

You'd probably guess that a love for Israel would be high on the list. Religion too – where better to live a Jewish life than in the ancient homeland of King David and our biblical heroes? There are also strong economic incentives: as expensive as Israel is, the country offers financial benefits and discounts to those who make aliyah, such as a free master's degree to *olim* under age thirty, tax waivers on foreign income, and incredibly affordable health care.

According to Marc Rosenberg, however, these reasons, while all perks, are not why people come. And he would know, being the Jerusalem-based vice president of Diaspora relations for Nefesh B'Nefesh, the organization that has helped more than seventy-five thousand people make aliyah since 2002. And if there's one thing he knows after all those years on the job, it's this: people today don't make aliyah solely out of a love for Israel or Judaism, or for other ideological reasons. As Rosenberg explains, "My brother has the exact same ideology and love for Israel as I do, and he lives in the US."

The difference, according to Rosenberg, has less to do with Israel and more to do with the person. "I'm in Israel because I was more willing to take a risk and step out of my comfort zone," he says. "That's the one characteristic I find in almost all *olim* – that willingness to take the plunge."

This ability to buck convention and take a huge, life-altering risk is, in Rosenberg's opinion, why making aliyah is such a daunting experience for all *olim*, especially from the States: it necessarily requires a person to take a bold action that goes against the grain. "Americans in particular like to plan," Rosenberg says. Those of us from the US can no doubt relate to this; take, for example, the pressure to choose a good nursery school for our kids so that they'll be reading by kindergarten, so that they can be in advanced classes in elementary school, then AP classes in high school, so that they can get into an Ivy League university...Sound familiar?

Rosenberg compares the "normal" American life path to a ski lift: there are certain stops to get off, but if you're between stops, the only way off is to jump. "Americans are afraid to jump, culturally," he says. "The ones who move here are the ones unafraid to take that leap."

This also explains why chutzpah is and probably always will be front and center in the Israeli ethos: between the original pioneers and today's *olim*, Israelis are a self-selected group of risk takers. Ben Gurion Airport might as well display a sign that says, "Abandon Shyness, All Who Enter Here." Good luck making it in the Middle East if you can't speak your mind and advocate for yourself.

"Jews have always immigrated for multiple reasons," says Dr. Avinoam Patt, director of Jewish Studies at the University of Connecticut. "Economic opportunity, religious freedom, to escape persecution, and on the basis of political ideologies like Zionism. But above all," he continues, "Jews have sought new homelands where they could build a better future and have the freedom to continue living as Jews." For *olim*, Israel is where they can live a Jewish life, feel that they're making an impact, and enjoy the energy and peoplehood of life in the world's only Jewish state.

David Mencer, a communications and fundraising specialist, made aliyah in 2017 from London with his wife and three children. "I had experienced the best that London had to offer," Mencer says. "The house, the cars, the job, the schools. I tasted it all and believed there was something better. Shimon Peres said that the Jewish people's greatest contribution to the world is dissatisfaction. It's easy to complain about the reasons to not come, but for me? It was an easy decision."

Harper Spero made aliyah in 2021 during the heart of the pandemic. For her, finding the courage to "jump" was nothing new. As a professional business coach, she has built a career on helping others overcome their own fears. "I've been self-employed, started businesses, launched podcasts . . . I have confidence in trying things out and seeing how they go. If it doesn't work, it doesn't work!"

For most people, the aliyah application process isn't complicated: forms, proof of Judaism, an interview. Selling one's house and cars, packing and shipping, and the logistics of the move itself are much harder. Still, the toughest part of making aliyah may just be *deciding to make aliyah*.

"A lot of prospective *olim* tell me they want to move to Israel but they're 'waiting for the right time,'" says Rosenberg. "The thing is, there *is* no right time. If you wait till you're ready, the time will probably never come."

Rosenberg's advice? "Visit. Research. Talk to other *olim* and hear what it's like. And then…"

He smiles. "Jump."

The Five Stages of Aliyah

Remember that high school crush who gave you butterflies? The college flame whose name you doodled in your notebook (or, to those who never lived without the internet, stalked on Instagram)? The first weeks of any romance are exhilarating, intoxicating, a nonstop thrill ride of electrifying emotions, and – as any married person will tell you – unsustainable. Because no matter how beautiful, funny, or smart your new love is, sooner or later you realize that this incredibly awesome person is actually just…a person.

"He/she just isn't who I thought he/she was" is one oft-heard explanation for breakups. But there's another group who knows this concept well: *olim chadashim*. Except in their case, their "partner" is Israel.

"One reason so many *olim* struggle is not because they don't love Israel, but that they love it the wrong way," says Jerusalem-based therapist Nomi Raz. For three years, Raz worked in Washington, DC, as a *shlichah* (emissary), helping prospective American immigrants navigate the logistics of moving to Israel. Since returning to Israel, she provides counseling for *olim* struggling to adjust.

As Raz notes, immigrants' connection with Israel often begins like a relationship with a romantic partner: worship, idealization, and putting the country on a pedestal. "To settle down and build a life, though, the immigrant needs to progress beyond infatuation," she says. Only by ditching the rose-colored glasses can they love Israel for what it is, not what they want it to be. Because as any immigrant will tell you, the spark will inevitably dim once you encounter the nuts and bolts of daily life in this hot, crowded, loud, expensive Middle Eastern country.

"When speaking with clients, I like to compare the aliyah process to a romantic comedy," Raz says. "The movie ends with a wedding, and we imagine the couple lives happily ever after. But as all married couples know, the wedding isn't the finish line, but the start of a new life." Many immigrants assume that the hardest part is the move itself, akin to planning the wedding – the paperwork, quitting your job, the packing, shipping, selling your house and cars, finding an apartment, and all the stressful logistics involved in the big move.

But as tedious as all of that is, the real challenge is what happens next: *klitah* (absorption) – the part the movie doesn't show.

Raz has seen this story enough times to recognize the pattern of "five stages of aliyah," a process of adjustment she learned from her colleague cross-cultural trainer Lucy Shahar.

1. **Euphoria.** "The bus driver says '*Shabbat shalom*' to me, the hummus is delicious, and I can wear sandals year-round! Israel is *sababa* [awesome]!"

2. **Depression.** "My kids don't understand a thing in school, and my Hebrew sucks, but how can I go to ulpan when I have to work? And customer service is *so frustrating* – Israelis are all so rude. I never should have moved here."

3. **Adjustment.** "I just signed a lease on a pretty nice apartment. It's definitely smaller than the house we used to live in, but it seems okay. My job is good...I wish the salary were higher, but we're doing alright. I've made a nice group of friends through shul and yoga class, and I can even pay my bills in Hebrew by myself! I can't wait for my parents to visit and see what I've accomplished."

4. **Disillusionment.** "My kids' education doesn't compare to what they'd receive in the US. The teachers seem more like babysitters. My friends back home are buying vacation houses, and I question whether we'll ever be able to own a home. Our children are growing up not knowing their grandparents. If we're ever going to move back, now is the time, before we send them to a new school. Everything is fine, but I kind of expected more...Is this as good as it gets?"

5. **Biculturalism** (acceptance). "I am at peace with the fact that wherever I live, something is missing. Here, I feel American; there,

I feel Israeli. I've given up on having the perfect Hebrew accent, but who cares? Everyone understands me and that's what matters. Sure, most of my friends are *olim*, but is that really so terrible? Most of my friends in the US were Jews. In some ways, I'll always be an immigrant, and I'm fine with it. It may not be easy, but it's never dull. As long as the positives outweigh the negatives, what else can I ask for?"

An immigrant who can make it to that last stage of successfully navigating two different cultures, Raz says, will be able to build a long and sustainable life here. And if you don't make it, she thinks it's unfortunate. "Not because I believe all Jews should live in Israel," she says. "But if you leave too early, at the heart of depression, how can you know how you'll feel once you get to the other side?" Raz once again draws the comparison to marriage: it's normal to wonder what "could have been" or what else is out there. But only by making a good-faith effort to work through the obstacles can you know if you made a decision you can live with. And if you do choose to split up – or leave Israel – you can do it with a full heart and inner peace.

Now, what exactly *is* the secret to a successful aliyah?

For that answer, we turn to another expert. The great Carl Jung claimed that when we idealize our relationship partner, we are really projecting – meaning, we see people as we'd *like* to see them, rather than who they truly are. To apply the metaphor to aliyah, imagine a couple who moves to Israel after experiencing only the highs of the country and the magical feeling of "coming home" that so many people have on their visits here. What this couple may not realize is that a vacation, *especially to Israel*, is not real life.

No matter how deeply you love Israel while watching the sunrise over Masada, floating in the Dead Sea, or praying with your Jewish brothers and sisters at the Kotel, those activities have little in common with those that fill an Israeli's average day. Real life is more likely to involve activities such as having to visit your old bank branch in a different city in order to complete a form because the Israeli bank system isn't centralized, racing to the pharmacy on Friday afternoon before it closes for Shabbat, or taking your kids to work because there's another teacher strike...

Will that couple still love Israel after vacation ends and real life begins?

Whether or not the move works out depends on the immigrant's ability to know the *real* Israel, warts and all. Marc Rosenberg of Nefesh B'Nefesh offers this advice: "Spend time here beforehand," he says. "A summer job, a gap year, an internship. Not as a tourist on a bus, but as someone who actually lives here." He explains that if you discover the hard parts and still want to live here, you're less likely to be surprised and discouraged later.

And while there's no study to back it up, Rosenberg has found that in addition to a willingness to take risks, the other quality common to successful *olim* is a high tolerance for ambiguity. "There's a good chance that in their home country, the immigrant felt empowered with a high sense of control," he says. "Once they arrive in Israel, that sense of control will often vanish. To successfully acclimate to life here – or any new place – the immigrant needs to be able to say, 'I have no idea what's going on right now, I don't know what's going to happen next, I can't control any of this...

"'And I'm okay with that!'"

Cross-Cultural Dating: "It's Your Loss!"

Negotiations filled with misunderstanding. Parties not on the same page. Grievances and emotional needs unmet.

Are we talking about peace talks? Not even close. More like the challenges new immigrants face when dating Israelis.

During a first visit to Israel, the average Jewish tourist will fall in love with locals multiple times, drunk on the novelty and allure of all these gorgeous, exotic Jews. Is the typical Israeli really better looking than the beauties of, say, Toronto or Paris? Maybe, maybe not. But, as Jews, we *believe* they are, because they're Jewish – and for whatever reason, this makes them more attractive.

For this reason, tourists often think, "Why, if only I lived *here*, surrounded by all these beautiful Jews, I'd have a date every night of the week!"

And that's where the trouble starts. Young immigrants figure dating Israelis will be easy, now that the question of Jewishness is out of the way; in fact, some young Jews immigrate precisely *because* they want to find a mate. What the new immigrant often fails to take into account, however, is that a new question has taken the place of "Is he/she Jewish?": how to overcome the cultural gap between immigrants and sabra Israelis.

To get the inside scoop, we spoke with two dozen single *olim chadashim* about the adventure of dating Israelis.

To help us make sense of the differences we uncovered, we turn to professional matchmaking team Ravit Migdal and Billy Scheingart.

The Challenges

"We still on for five minutes from now?" Whereas seasoned Anglo daters expect to make weekend plans *before the weekend*, Israelis like to finalize plans at the last minute. Though even the word *plans* might be an exaggeration, because that implies some sort of, you know, *plan*. More often, the furthest things get is "We should get together." This leaves many new immigrants – especially women – waiting by the phone. A woman may not know if she should accept other offers because the guy who asked for her number on Sunday still hasn't been in touch.

Complicating things further is that even on a weeknight, no hour is too late to meet – so don't be surprised by an unexpected 10 p.m. text message that says, "*Yalla*, what are we doing?" To a Western Anglo, this may sound inconsiderate and even rude, but matchmaker Ravit says otherwise.

"Israelis live in a state of constant uncertainty," she explains. "Deep down, we feel we can never know for sure what's going to happen tomorrow, or if tomorrow will even come. 'What if a war starts? What if the Nazis attack? I could *die* tomorrow, so talk to me tomorrow evening, and if I'm still alive, we'll go out – if I still feel like going out with you.'"

"No picnic basket, no mini golf? What gives?" Abroad, many consider a creative first date idea to be romantic; in Israel, though, "going the extra mile" is just creepy. Any unfamiliar behavior or proposal that goes against the grain is considered a turnoff. The typical first date in Israel is relatively tame, perhaps even cliché – often nothing more than a meet-up for coffee or a glass of wine. According to Billy: "When everyone lives in a state of survival, and just about everyone has been through the army, including women, this is what happens. We're constantly on alert, and anything non-standard sets off alarm bells."

Women don't trust men. As one well-mannered American guy we spoke to put it: "Israeli men ruined it for the rest of us." The machismo and aggressiveness of Israeli men have conditioned Israeli women to keep their guard up and set boundaries that will be breached if and only when they decide. Likewise, Israeli women often don't know what to make of men who don't fit the classic mold of the Israeli male, no matter how good their intentions; indeed, Hebrew lacks a word for "chivalry" or "gentleman."

Another immigrant male relates this story: "On the third date, I offered to cook a romantic dinner at my apartment. She said no, because to her, dinner meant sex." In other words, to the Israeli woman, there's no such thing as nice just for the sake of being nice. As Ravit explains: "All women were in the army, and we know how to protect ourselves. We're not naive, we don't fall for tricks, and you're going to play by our rules."

"Good luck finding someone better than me!" More than one immigrant we spoke to noted that after they broke off the relationship, their sabra ex texted back "It's your loss," "You'll regret it," "Good luck finding someone else as good as me," or some other face-saving message. Nobody likes to get hurt, but in the Middle East, Billy says, both men and women are more likely to act in ways that protect their egos.

The Perk

Getting to know you, fast. Several of the people we talked to, of both sexes, noted that first-date conversation is more real and meaningful than in other countries. It's common for a new couple to split a bottle of wine and talk deeply about the most intimate and even painful experiences of their lives. And if there's a big holiday coming up, such as Rosh Hashanah or Passover, don't be surprised if your sabra date invites you to their parents' house for the holiday dinner, no strings attached. It's a genuine invite; embedded in Israeli culture is the idea that when the holidays roll around, anyone without family in Israel should have a place to go.

The "It Depends How You Feel about It"

"Too bad you didn't like me. How about a kiss?" We've been told that Israeli men like to try for a kiss, no matter what. For some women, this is a positive, but others are uncomfortable with this, as it puts them in the awkward position of having to push the guy away or sneak off with a quick goodbye. "Israeli men are bred to be aggressive," says Ravit. "Part of it is the army, part is that this is, after all, the Middle East, and some of the dating rituals are a bit primitive." One single Australian woman put it differently: "From the guy's point of view, the goodnight kiss is simply a fifty-fifty gamble – the woman will either like it or she won't, so he's got nothing to lose."

The Concession

The sky's the limit. As one young Canadian single told us, "Israeli men think they're God's gift to women – probably because their mothers spoil them from a young age, knowing they'll one day go to the army. And fighter pilots? It's that times a thousand. Their egos are enormous.

"Though I gotta admit," she adds, "they're incredibly hot."

"What Did I Just Say?": Embarrassing Hebrew Mistakes

A stranger in a strange land is bound to make mistakes. For many *olim*, no gaffe is more embarrassing than the linguistic slip-up.

Don't worry, says Aviv Bertele, founder of the popular Tel Aviv ulpan (an intensive Hebrew instruction center) that bears his name. Israelis mostly find our mistakes cute and endearing. And they appreciate the effort, taking into account that very few people on the planet even speak Hebrew. "Israelis can be very judgmental, but when it comes to Hebrew, they're actually much more tolerant than most people from other countries," Bertele reassures us. "When they say '*Kol hakavod!*' (Great job!), they're not being sarcastic. They mean it!"

According to Aviv, some of the most common mistakes come from words that share the same *binyan* (verb construction) and a similar vowel sound – words like *holech* (to walk or go), *ochel* (to eat), and *ohev* (to love). As you'll see in the examples below, a word that can sound right can be oh, so wrong! The good news is that no matter how embarrassing the mistake, you can be sure that you weren't the first to say it – or the last.

Below are some of our favorite English-to-Hebrew mix-ups, courtesy of a few dozen *olim* we reached out to. While the list concludes with some of the more … shall we say … "adult-themed" mistakes, we have omitted the most crass.

While Shopping…

What they meant to say	**What they actually said**
"Ani tzrichah liknot gluyot."	*"Ani tzrichah liknot glulot."*
"I need to buy postcards."	"I need to buy birth control pills."
"Rak prusah echad, b'vakashah."	*"Rak prutzah echad, b'vakashah."*
"Just one slice, please."	"Just one prostitute, please."

"Ani ekach challah metukah."
"I'll take a sweet challah."

"Ani ekach challah motek."
"I'll take a challah, sweetie."

"Efshar la'azor li? Ba'ali mishto'el."
"Can you help me? My husband is coughing."

"Efshar la'azor li? Ba'ali mishtolel."
"Can you help me? My husband is going wild."

"Ani mechapes michnesei chaki."
"I'm looking for khaki pants."

"Ani mechapes michnesei kaki."
"I'm looking for excrement pants."

When Dealing with Customer Service...

What they meant to say	What they actually said
"Ani me'unyan l'kabel halva'ah." "I'm interested in getting a loan."	"Ani me'unyan l'kabel levayah." "I'm interested in getting a funeral."
"Ani tzrichah mishehu l'chasel et hacharakim." "I need someone to get rid of the insects."	"Ani tzrichah mishehu l'chasel et hacharedim." "I need someone to get rid of the Haredim (ultra-Orthodox Jews)."
"Al titkasher kol kach mukdam machar – hayom adayin yashanti." "Don't call so early tomorrow – today I was still sleeping."	"Al titkasher kol kach mukdam machar – hayom adayin hishtanti." "Don't call so early tomorrow – today I was still urinating."
"Efshar la'azor li livchor mazgan? Kol kach cham li ba'bayit!" "Can you help me choose an air conditioner? It's so hot in my home!"	"Efshar la'azor li livchor mazgan? Ani kol kach cham ba'bayit!" "Can you help me choose an air conditioner? I'm so horny in my home!"

Explaining One's Behavior...

What they meant to say	What they actually said
"Slichah she'ani me'ucheret."	"Slichah she'ani mechu'eret."
"I apologize that I'm late."	"I apologize that I'm ugly."
"Bati l'vaker otach."	"Bati livkor otach."
"I came to visit you."	"I came to bury you."
"Ani lo yecholah l'hipagesh – ani tzrichah l'hitkaleach."	"Ani lo yecholah l'hipagesh – ani tzrichah l'hitlakeach."
"I'm unable to meet – I have to shower."	"I'm unable to meet – I have to burst into flames."

With a Complete Stranger...

What they meant to say	What they actually said
"Rega, nahag! Ani tzarich laredet!"	"Rega, nahag! Ani tzarich laledet!"
"Wait, driver! I need to get off!"	"Wait, driver! I need to give birth!"
"Efshar la'azor li? Halachti l'ibud."	"Efshar la'azor li? Hitabadti."
"Can you help me? I'm lost."	"Can you help me? I killed myself."
"Slichah, mah hasha'ah?"	"Slichah, ma haShoah?"
"Excuse me, what's the time?"	"Excuse me, what's the Holocaust?"
"Boker tov, korim li _____. Na'im me'od"	"Boker tov, korim li _____. Ta'im me'od"
"Good morning, my name is _____. Nice to meet you."	"Good morning, my name is _____. Very tasty."
"Echpat lecha l'lavot oti l'moniot sherut?"	"Echpat lecha l'lavot oti la'sherutim?"
"Do you mind accompanying me to the group taxis?"	"Do you mind accompanying me to the bathroom?"

At the Office...

What they meant to say	What they actually said
"Ani oseh to'ar b'kalkalah." "I'm doing a degree in economics."	*"Ani oseh to'ar b'economicah."* "I'm doing a degree in bleach."
"Ani afitz et hahoda'ah la'itonut b'rachavei ha'olam." "I am going to distribute this press release around the world."	*"Ani aflitz et hahoda'ah la'itonut b'rachavei ha'olam."* "I am going to fart this press release around the world."
"Ani tzrichah rechev l'hagia l'misrad." "I need a vehicle to get to work."	*"Ani tzricha rechem l'hagia l'misrad."* "I need a uterus to get to work."

To Avoid at All Costs...

What they meant to say	What they actually said
"Zeh mamash kasheh; efshar l'pashet?" "This is very difficult; can you please simplify it?"	*"Zeh mamash kasheh; efshar l'hitpashet?"* "This is very difficult; can you please take off your clothes?"
"Slichah! Lo zihiti otcha im mishkafayim!" "Sorry! I didn't recognize you with glasses!"	*"Slichah! Lo zihiti otcha im michnesayim!"* "Sorry! I didn't recognize you with pants!"
"Efshar mitz gezer?" "May I have some carrot juice?"	*"Efshar mitz gever?"* "May I have some man juice?"
"Eifoh hayita? Dafakti lecha b'delet!" "Where were you? I knocked on your door!"	*"Eifoh hayita? Dafakti otcha b'delet!"* "Where were you? I made love to you on the door!"

"*Kama oleh bul echad, lishloach
l'Anglia?*"
"How much does it cost for one
stamp, to England?"

"*Kama oleh bulbul echad, lishloach
l'Anglia?*"
"How much does it cost for one
penis, to England?"

"*Ani elech l'misrad shelo machar
v'avi b'yad.*"
"I'll go to his office tomorrow
and deliver it to him by hand."

"*Ani elech l'misrad shelo machar
v'avi lo b'yad.*"
"I'll go to his office tomorrow
and pleasure him by hand."

So if you do make aliyah or even come for a lengthy visit, we can't recommend learning Hebrew strongly enough. Whether it's Aviv Bertele's or any of the country's numerous ulpanim, even minimal Hebrew proficiency will dramatically enhance your Israel experience. And don't worry – no matter how difficult the process may be, we promise your teachers will make it as simple as possible.

Or at least help you take off your clothes.

"I Remember When…": Israel through the Decades

Do you remember when America's founding fathers signed the Declaration of Independence? When the Archbishop of Canterbury drafted the Magna Carta? When the French working class stormed the Bastille?

Unlikely. But the State of Israel is still young enough that some reading the first edition of this book remember where they were on May 14, 1948, when David Ben-Gurion stood in the Tel Aviv Museum of Art, today's Independence Hall, and proclaimed the establishment of the State of Israel. And even when that will no longer be true, you can still watch Ben-Gurion's speech on YouTube. Can't say that about George Washington.

Cherishing memory and preserving our history have always been integral parts of the Jewish people. In this chapter, we recall a few historical moments from throughout Israel's first seventy-five years.

We also listen to the memories of *olim* who experienced Israel's highs, lows, growth, history, and twists and turns firsthand.

The 1950s

- The Knesset moved from Tel Aviv to Jerusalem's King George Street.
- Egged merged with two other bus companies, creating a national transportation network.
- "I remember how all of a sudden we could make direct telephone calls anywhere in the country without having to be connected by an operator."

The 1960s

- Adolf Eichmann was put on trial and hanged.
- "I'll never forget the panic and fear before the Six-Day War. We heard they wanted to throw us into the sea. As a seventh-grader, I helped dig trenches throughout our village."
- "When I mailed a letter to my family in the US, it took at least a week to arrive, and then another two weeks to hear back. Once a year – yes, only once – I would make a phone call home, which required making an appointment at the central post office."

The 1970s

- Israel qualified for the World Cup for the first and only time.
- Rina Mor became the first and only Israeli to win the Miss Universe pageant.
- Israel won the Eurovision contest for the first time with "A-Ba-Ni-Bi" by Izhar Cohen.
- "My fellow *olim* and I had a system with our parents in our home countries: To let them know we were okay, we would make a collect call – which they wouldn't accept. If we actually wanted them to answer, we'd call right back."

- "To make a phone call, you needed an *asimon*, a silver token with a hole in the middle. We would tie a sewing thread around the coin, and a few seconds before our time ran out (we had a stopwatch of course), we would jiggle and pull the thread in short, quick movements. You'd hear the click indicating time's up, but the call would continue. You could talk to folks abroad for over twenty minutes with a single *asimon*!"
- "Inflation hit over 400 percent, and I needed to work two jobs. Every time we went to the store, the price of bread was different."
- "During the Yom Kippur War, we 'blacked out' towns so the enemy army couldn't see where they were aiming. We covered our windows with blinds and dark sheets, and kids from schools and youth movements would paint people's headlights black to minimize the light."
- "I grew up in a lovely house in the Sinai. Then Prime Minister Begin shook Anwar Sadat's hand on the White House lawn and made peace, with Israel giving all of Sinai to Egypt. My family was forced to move, and I had to say goodbye to all of my friends."

The 1980s

- The Egyptian embassy opened in Tel Aviv.
- The shekel replaced the lira as the national currency (which the new shekel later replaced).
- Simon and Garfunkel performed in Ramat Gan Stadium.
- "We had hyperinflation, so at the end of the month, your paycheck was worth less than the month before. Banks became more generous with overdrafts, unheard of in the US. People knew that if they bought a microwave, in a month's time, it was still worth a microwave, but if they waited, their 100,000 shekels would only be worth 90,000. It was absurd."

The 1990s

- One million Russian immigrants made aliyah.
- Dana International won Eurovision.

- On the front page of the newspaper the day after the Ason Hamasokim (Helicopter Disaster) in February 1997, there were seventy-three photos, one of each of the soldiers killed.

- "I remember watching TV when Yitzhak Rabin shook hands with Yasser Arafat on the White House lawn. My friends and I cried. We truly believed peace had come."

- "I made aliyah four months before Yitzhak Rabin was assassinated. The following year, there were bus and mall bombings nonstop in Jerusalem and Tel Aviv. That's when security guards were first placed at the entrance to malls and restaurants. The Four Mothers movement started protesting, and suddenly we all realized that having soldiers in Lebanon made no sense."

- "It was so much harder then to find other English-speaking immigrants like me. Now there's Facebook, social media, and Nefesh B'Nefesh, but back then I had to figure it all out by myself."

- "My entire adolescence took place on the backdrop of the Second Intifada. I was thirteen when it broke out in 2000, and I remember the worst days of 2002 all too clearly. I had just graduated high school when the 2005 disengagement from Gaza took place, and that was very significant, too. The sense of Israeli solidarity during times of trouble definitely helped me understand why my parents had uprooted us from our comfortable life in Pittsburgh to come here. Is that weird? I felt that I was part of something, that this was where I belonged and it would have been harder, as a committed Jew and a Zionist, to be watching all this happen from far away."

The 2000s

- "I came to Israel on the first Birthright Israel trip."

- "The conflict was heart-wrenching in the early 2000s when buses were blowing up. I was never sure I would see my husband at the end of the day. We used to say goodbye 'forever' in the morning before we left the house."

- "Before FaceTime, WhatsApp, and Zoom, there was Skype. I bought my first laptop weeks before I made aliyah with – are you sitting down? – a built-in camera. My family didn't have webcams, but by paying five dollars a month, I was able to get a virtual phone number they could call to reach me on my computer. Of course, we had to schedule the calls so I could be available."

- "I arrived during the Second Lebanon War in 2006, one year after the disengagement from Gaza. No one could understand why I moved, neither my Australian friends and family nor Israelis. There seemed to be a massive chasm between how the situation was reported and discussed in Israel and how it was portrayed in foreign media."

- "I had just moved here, and it was so hard for my family. It was impossible for them to imagine or relate to what day-to-day life was like when things heated up. When there was some kind of terrorist event, I'd email them immediately to let them know I was okay. Now, with video, quick communication, and all of us simply being 'used to it,' they know I'm as safe here as anywhere else in the world."

- "Gilad Shalit was kidnapped by Hamas, and they wouldn't release him. Life continued here for the rest of us, but every so often, I would find myself thinking about him, that right now Gilad Shalit is *there*, in Gaza probably, held by Hamas, and here I am buying a cappuccino."

The 2010s

- Gilad Shalit returned home after five years in captivity.
- Google purchased Israeli start-up Waze for over a billion dollars.
- The American embassy moved to Jerusalem.
- "Operation Protective Edge in Gaza happened in late summer. Every hour, when they read the news, I both dreaded listening and couldn't turn away. Sometimes there would be nothing new to report, and that was a relief, but too often they would announce that soldiers were killed, and then I would literally break down crying, thinking about the mothers and fathers who one minute

had been cooking dinner or ironing their clothes and a moment later their lives were destroyed."

- "I spent 4,200 shekels a month on rent for my first apartment in the Nachlaot neighborhood of Jerusalem. Every time it rained outside, it would rain in my bedroom. When we showered, the water flooded into the kitchen. When we got mold, the landlord's solution was to pay someone to apply plaster and paint so we wouldn't see it. In our new place, we have to wait two hours for hot water, meaning that taking a shower can be a day-long commitment. Living in Israel has definitely made me not spoiled. When I have instant hot water in the States, I'm going to be in shock."

- "I moved to Israel just before the outbreak of COVID-19. The entire country shut down one week after I arrived! At least the government gave us a small 'sorry about the pandemic' grant."

The 2020s

- "I talk to my family via FaceTime, WhatsApp, and Instagram, and we do a monthly Zoom call for the whole family."

- "The conflict hasn't affected my life at all. I'm surprised how little it is discussed among my friends. I think it's because the rest of the world feels so out of control right now that this is the 'norm' here."

- "I had to fight my way into getting an appointment just to open a bank account. You can't just show up at the bank and say, 'I want to give you my money' and then when you do get an account open, they put a 5,000-shekel monthly spending limit on my credit card. Not helpful when you're trying to set up your new life!"

- "When Operation Breaking Dawn began, the idea of running to the bomb shelter had me terrified. But my more experienced immigrant friends weren't even phased. They'd say, 'You'll get used to it. It's always either this or elections.'"

- "Everyone I talk to thinks civilization is falling apart these days. Climate change, nationalism, COVID ... If the world is going to end, there's nowhere I'd rather be than right here."

I Knew I Was Israeli When…

An *oleh chadash* who arrives in Israel is immediately ushered into a Ministry of the Interior office at Ben Gurion Airport, handed a *te'udat oleh* (immigrant ID card) and a temporary *te'udat zehut* (national ID card), and more importantly, becomes officially Israeli.

Feeling like a real Israeli, however, is an entirely different matter. Even immigrants who grew up with one or two Israeli parents, speaking Hebrew at home and visiting relatives in Israel every year, struggle to acclimate. Over and over, immigrants mention that adjusting to the culture is one of the biggest challenges.

That said, there are moments when most new immigrants begin to realize that they're more Israeli than they thought – and less the outsiders they used to be.

We reached out to dozens of new immigrants for the moments or experiences that completed the sentence "**I knew I was Israeli when…**"

1. I pulled over my car and urinated on the side of the road.
2. I asked to go ahead of someone in line at the grocery store, since I only had a handful of items.
3. My roommate told me I had spoken in Hebrew while dreaming.
4. I ate a whole raw onion at a *hummusiah* (hummus restaurant).
5. I booked a post-army Pesach trip to Thailand for my son.
6. I no longer realized when it was Christmas, Labor Day, or Memorial Day back in the US.
7. I was in the mood for peanut butter and jelly, had none at home, and instead used raw tahini and mashed date spread.
8. I pronounced the English curse word as "*sheet.*"
9. On Sunday morning when visiting the US, I was astonished that nobody went to work.
10. I remarked to my Israeli husband that the radio announcer had made a grammatical mistake in Hebrew.
11. I yelled at the security person at the airport.

12. I went to the US consulate to renew my passport on July 4 without realizing it would be closed.

13. I got excited when it rained.

14. I switched from Taster's Choice to Turkish coffee.

15. I actually used the horn in my car.

16. I argued with my boss without feeling anxious about it.

17. I started saying *"Eemaleh!"* (Mommy!) instead of "OMG!"

18. I found out that a stabbing attack had occurred where I had been the day before, just ten minutes from home.

19. I visited New York City, and everyone seemed so polite.

20. I texted people a question without first saying, "Hi, good morning."

21. I let my shopping cart hold my place in line at the supermarket while I did my shopping.

22. I started saying "Ehhhh" instead of "Uhhhh."

23. Back in the US, I referred to 7-Eleven as "the *makolet*" (neighborhood grocery store).

24. I called a condescending service provider *"Mami"* (my dear) in a snarky voice.

25. I had a zero balance in my checking account but was just happy that I wasn't in *meenus*.

26. I did the *"Rega!"* (Hold on a minute!) hand gesture (see "More than Words: Nonverbal Communication" in chapter 3).

27. I heard Israeli songs on the radio and became nostalgic for different times of my life in Israel.

28. I drove in the States and honked at someone for not moving the moment the light turned green.

29. I voted in an Israeli election for the first time. And then soon after, the second time.

30. I became a hummus snob, refusing to eat anything packaged and sold in the grocery store.

31. I passed a car by driving on the shoulder of the highway.

32. I visited America and made people uncomfortable by how close I stood to them, without realizing it.

33. Instead of speaking two languages fluently, I just became bad at both.

34. I elbowed my way onto the bus ahead of others in Canada.

35. I attended the swearing-in ceremony of my sabra son as he became an officer in the IDF.

36. I yelled at a cashier in America, and everyone backed away from me.

37. I jumped out of a moving car in the US to stand in a parking spot before someone else snatched it.

38. I said "*pop-kor-en*" instead of "popcorn."

39. I was told in the US that I have a big mouth and an opinion on everything.

40. I wished the Arab delivery driver the holiday greeting "*Ramadan kareem.*"

41. At London's Finsbury Park train station, I heard two people chatting in English and automatically turned around to see if I knew them.

42. I realized my childhood friends from America could hardly understand simple English sentences: "*Davka* [actually] it's not *kedai* [worthwhile] to buy it here – you can get a *mivtzah* [deal] in Rami Levy or the *shuk.*"

43. I traveled to Thailand for vacation.

44. Back in England, I went to shul and was struck by what a strong English accent the cantor had when he davens.

45. I said something correctly in the grammatical construction *hufal*.

46. I shoved my way through a crowd, pulling my dad behind me to get gas masks for him and my mom.

47. I made peace with the fact that the weather report tells you that it's either going to be hotter than usual or colder than usual and nothing more.

48. Back in America, I parked on the sidewalk.

49. I wrote the wedding card blessing on the outside of the envelope.

50. I traveled to India and did "*shvil hahummus*" ("the hummus trail," a popular backpacking route for Israelis just out of the army).

51. I could quote famous lines from the movie *Givat Halfon* at the appropriate moment.

52. I stopped doing the quick mental math to convert prices from dollars to shekels.

53. I stopped showing up for social engagements on time; instead, I would arrive two hours after the originally proposed time and was still the first one there.

54. I stopped moving aside when people walked toward me in the other direction.

55. I dressed up for Purim as a character from an Arik Einstein song.

56. I stopped saying "*slichah*" (excuse me) when bumping into people.

57. When I entered an office and wasn't offered coffee, I asked for it.

58. Commuting more than an hour to work became unacceptable.

59. I yelled at the support rep that his service was "*al hapanim*" (terrible).

60. I stopped converting Celsius to Fahrenheit.

61. When someone cut in front of me in the checkout line, I started screaming at him in Hebrew. This was in Idaho.

62. I could no longer remember the last time I wore a tie.

63. I heard a sonic boom overhead and it didn't faze me.

64. It stopped bothering me that my child said "Nope" when I asked if she had any homework.

65. I hailed a taxi and sat in the front seat because it was easier to chat with the driver.

66. I asked the teenage waiter if she'd please watch my baby while I went to the restroom.

67. I cracked a joke in Hebrew, and people laughed.

68. My friend in the States complained about paying four dollars for a gallon of gas, and I thought, "Wow, what a bargain!"

69. I realized I preferred drinking water at room temperature, not served on ice.

70. I stopped waiting until I went abroad to buy my deodorant for the year.

71. I caught myself making fun of tourists for seeming "so American."

72. I entered Gaza with my IDF infantry platoon.

73. In the US, someone asked me a question and instead of answering "No," I just clicked my tongue.

74. I dropped off my visiting American friend at Ben Gurion Airport and drove off, so happy that I got to stay.
75. I felt complete.

Face to Face with Professor Noah Efron

PERHAPS NO OTHER YEAR IN ISRAEL'S HISTORY AFFECTED THE Israeli psyche more than 2023. On January 4, newly appointed justice minister Yariv Levin declared the government's intent to reform the judicial system. Months of tense protests over legislation and checks and balances quickly evolved into larger fights over the deeper question of "What does it mean to be a Jewish state?" By the end of the year, however, the country's focus had unexpectedly turned to external threats after the shocking October 7th attacks by Hamas.

Professor Noah Efron is a published author, chair of Bar-Ilan University's interdisciplinary program on Science, Technology, and Society, cohost of TLV1's *The Promised Podcast*, a founder of Israel's Green Party, and has served on the Tel Aviv-Jaffa City Council. We spoke with Professor Efron to better understand both the short- and long-term effects of this year on the country and its mindset.

Benji Lovitt: It's been said that after the October 7th attacks, the country will never be the same. Is this true, and, if so, how?

Noah Efron: The country is always "never the same." There are two Greek philosophers – Heraclitus, who said that "all is change; you can never step in the same river twice," and Parmenides, who said

252 · ISRAEL 201

that "change is an illusion" – and when it comes to Israel, they're both right.

Meaning?

We no longer have the illusion that we are firmly in control of our own defense, that our army can't be second-guessed. That our abilities to create brilliant new technologies are enough to protect us, or that the equilibrium with the Palestinians could continue ad infinitum. At the same time, I don't think we're a fundamentally different country.

How have we *not* changed?

We actually now see better who we always were and how deep the Israeli character runs. The outpouring of goodwill, the reservists flying home from abroad, the creative ways to serve the country ... in the past, we imagined these actions to be deeply Israeli, but after October 7th, our hopes were confirmed that this is, in fact, who we are.

What about our notion that Israel was the one place where Jews would always be safe?

I felt safe growing up in Silver Spring, Maryland. I'm not sure I ever felt 100 percent safe in Israel. Is anyone completely safe *anywhere*? That said, after hundreds of years of persecution, pogroms, and trying to appeal to the world for help, and despite the monumental failures of October 7th, Israel remains the one Jewish state with a Jewish army. That was the dream of Zionism. If you want to live as a Jew where the people responsible for ensuring your safety are also Jews, Israel is the only place to be.

October 7th also could have been much worse: Hamas called upon Palestinian citizens of Israel to rise up and join the fight, but they didn't. What can we learn about our vision of a Jewish state with such a large non-Jewish population?

That Palestinian-Israelis are truly citizens of Israel who feel deep ties to the country, despite it being emotionally and politically complicated for them. I spoke with Palestinian friends who were afraid to speak

Arabic in public or post certain ideas on social media. I now further appreciate that expressing solidarity with their fellow Israelis while being desperately worried about their loved ones in Gaza required quite the delicate choreography, at a very tense and frightening time. Eventually, I think we'll take more seriously that Palestinian citizens of Israel can actually be a bridge between the two peoples.

It's easy to forget that leading up to October 7th, deep internal rifts had appeared amongst the Israeli people. Have these rifts disappeared or simply gone into remission?

We Israelis still have profoundly different visions about what the country should be. Soon enough, it may be that we'll return to fighting and questioning each other's worldviews, no less angry and partisan than before; however, we now realize that our internal divisions are not existential. Because now that we appreciate each other's commitment to this project, we don't need to fear that the greatest experiment in Jewish history might fall apart.

But how much can one country take? The emotional toll of these back-to-back crises isn't too much to handle?

For all the unending tragedies of this and every other war, it's important to recognize that after the greatest tragedy that most of us have likely ever seen, we've also witnessed the most extraordinary expression of human solidarity, love, creativity, power, and mutual concern that we'll ever see. If someone had told us that thousands of people would put everything aside for weeks and months at a time, opening up their homes to refugees, cooking and delivering meals for entire platoons of soldiers, people would have said, "That's yesterday's Israel, not today's."

Finally, in such a bleak time, what about the country and its people gives us reasons to be optimistic for the future?

While things are mournful, they're not bleak. As horrific as October 7th was, I see the 8th of October as the true turning point of our lives. We saw with complete clarity what was always there: this solidarity, this outpouring of goodwill and energy, the willingness to sacrifice, and the belief that when you look left and right, the people alongside you are part of a common cause. Whether they have long *peyot* (religious side curls) or are speaking Arabic, they are beautiful, committed, energetic, creative, powerful people who believe in this project of the State of Israel. That's not bleak, it's profound and beautiful.

Face to Face with Futurist David Passig

THROUGHOUT *ISRAEL 201*, WE'VE TAKEN A JOURNEY INTO THE heart and soul of this magical, mysterious, and often chaotic country. But rather than focus on the themes typically found in a beginner's course or first-timer's itinerary, we introduced you to a variety of characters, institutions, phenomena, and issues in hopes of exploring Israel on a deeper level. As our journey comes to an end, our hope is that you have a new appreciation for Israel and that you leave with more questions than answers.

One such question might be this: What comes next? Or, what will Israel look like fifty years from now? A hundred?

No one knows for sure. But if anyone might, it's Dr. David Passig – Bar-Ilan University professor, bestselling author, and award-winning futurist.

Unlike historians, who concern themselves with the past, Professor David Passig analyzes current and past political, religious, social, and technological trends to predict the future. Among other watershed events, Dr. Passig correctly predicted the September 11 attacks and 2008 financial crisis, as well as the advent of mobile phones and "smart" AI-based appliances.

Passig is the author of numerous books, including *The Future Code* and *2048*, both of which won Israel's Gold Prize, the country's highest award for literary achievement. His most recent, *The Fifth Fiasco*, explores nothing less than the future and fate of the Jewish people.

Joel Chasnoff: In your lectures, you talk about the importance of what you call a "national ethos" and how Israel hasn't yet found one. What is an ethos, and why is it important?

David Passig: Throughout history, all great nations have had an ethos – an underlying principle that defines who they are and communicates to the world how and why their nation is different from others. For France, the ethos is "fraternity," and in America, it's "personal freedom." But Israel still doesn't have one, and for that reason the country is at a crossroads. To survive as a nation, we must define who we are and what we stand for. But the task has proven difficult.

Why?

Because this is new for us! For thousands of years of Jewish history, we've lived as tribes, in kingdoms, or as a minority in others' lands, but never as a sovereign nation of our own. So we're still not quite sure who we are. Since our establishment, we have defined ourselves as "a Jewish state," but what does that mean? You might say, "a nation that lives according to *halachah* [Jewish law]," but halachic Judaism is a product of being exiled from our homeland and scattered about the earth. We have a certain concept of what Judaism means, but the model we're familiar with today goes hand in hand with the circumstances we faced. It may very well be that our traditional concept of Judaism will continue to work, but we shouldn't assume that the model that served us in exile, scattered throughout the world, will be the best going forward – especially when more than 20 percent of our population is Muslim and Christian. They're Israeli too! Others have suggested we define ourselves as a "start-up nation," but guess what? Many countries want to be a start-up nation, so that doesn't distinguish us. Hence the challenge.

I want to return to a word you used earlier, "survival." Why does Israel's survival depend on finding this ethos?

Legitimacy. If we can't communicate to the world why the Jewish people need a country of our own and what that country stands for, we leave ourselves vulnerable to any and all who wish to delegitimize us. You see this today – when various groups try to tear us down, they hone in on that point, questioning our legitimacy.

I should add that what we're going through is typical of many new nations, particularly those born out of a cruel past. There's a pattern: For three or four generations, the various groups and sects put their differences aside for the purpose of building a new entity, one that ensures that history will not repeat itself. But then, once the new nation finds its balance, the various groups revert back to their original vision for what the country might be, and when these visions clash, the result is civil war. Not necessarily like the American Civil War, with hundreds of thousands of dead; it could be just one or a series of small attacks with thirty people killed. And it always starts with the assassination of a leader. After the shock of this event, people say, "Wait! Who are we? What are we doing?" And then the next generation creates a new idea around which the nation can reorganize.

I immediately think of the assassination of Yitzhak Rabin.

Exactly. That's where we saw the seeds of this process, three generations in.

Where are we now?

We are now in the fourth-generation era, when the next groundbreaking Jewish idea will be created. For thousands of years, we expressed Jewishness through ritual, prayer, and organizing ourselves into small, tight-knit communities, and time after time, we were slaughtered. Then came the most important Jewish idea of the past two hundred years: Zionism. The paradigm shift was dramatic. Zionism said, "Let's stop thinking about the Jewish religious narrative, and instead be proactive and use a modern tool – diplomacy – to break the cycle of our history and establish a sovereign state of our own."

Most Jews a hundred and fifty years ago didn't think it would work, and many opposed it, thinking it would be dangerous for us to congregate in one place. Thankfully, we've proven them wrong.

How so?

Just look at what we've accomplished. In under a century – less than a hundred years! – we have transformed ourselves into a thriving democratic state. We are a regional superpower, economically advanced, creating and sharing ideas that impact the entire world. Most important of all is that right now, the largest Jewish community in the world resides here, and it will only continue. By the year 2050, 70 percent of Jews will reside in Israel, and by the end of the twenty-first century, twenty million Jews will live here, accounting for 80 percent of world Jewry, with another four million Jews somewhere else.

Think about the enormity of that. In just two centuries, we're going to reverse the trajectory of Jewish history of the last two millennia! And we are part of that. This moment we are living in right now is one of the most significant in Jewish history, only we can't see it because we are in it. We talk about Pax Romana, but do you think the Romans knew they were living through Pax Romana? In the Middle Ages, did people know they were in the Middle Ages? It's the same for us, now.

You sound so optimistic, but I do have to ask: What about our enemies? Are you worried about Hamas, Hezbollah, Islamic Jihad? A nuclear Iran?

I'm not worried. Don't get me wrong, they pose a serious threat. But thank God, we have the means to take care of them if we need to. In the arc of Jewish history, these are blips on the radar.

You believe this even after the massacre of October 7, 2023?

I do.

To be clear, October 7th was an extremely painful moment for all of us. But in the arc of Jewish history, it won't have nearly as great an effect as people perceive. In the long run, there are more important and significant issues facing us, like our need to redefine exactly what it means to be a Jewish state.

What, then, is the correct way to think about the October 7th attack and its aftermath?

At the moment, just four months out, it's too soon to fully evaluate October 7th without the emotional fog that surrounds it.

But one thing I can say is that the October 7th attack and the war that followed and continues now have reminded us Israelis of our most important asset: our incredible ability to put aside our differences and unite as a people in pursuit of a common goal.

Our shared fate is the glue that binds all of us together. Building a Jewish homeland required us to put aside the many differences between us – Right, Left, Reform, Orthodox, communist, conservative, etc. – and channel our energies into making the dream of Israel a reality. October 7th reminded us that this bond is still relevant. It gave us a chance to redefine our raison d'être – because our fate is not yet sealed.

What you also need to understand is this: there has never been a time in history – anywhere, ever – when it was all peace or all war. We need to pan back and see this incredible moment for what it is.

It's a matter of perspective.

Yes. When you buy a coffee on Dizengoff Street, you don't think about the significance of that seemingly tiny act. We are the first generation of Jews who are truly free, to do what we want and go where we want in the world, all while knowing there is this larger entity that will take care of us. This change in mindset has freed us to think more boldly and creatively than we ever could. At every point in Jewish history, we have always required a plan B, with our suitcase packed and ready by the door. Even if we didn't say it out loud, we were thinking it.

But no longer. For the first time in our history in the Diaspora, there is no need for plan B.

Because there is Israel.

Yes! But it's even more than that. For the first time in thousands of years, the core of Jewishness is here. The profound debates about who we are and what we as a people are meant to be aren't happening anywhere else. That's why it's so hard to live here: because you can't not care. You *have to* care, you have to take sides. It isn't possible to disengage.

But that's also what makes this moment so incredible. For the first time, Jews can walk around with their heads held high and a sense of pride in who we are. *Israel has taken center stage in the unfolding of Jewish history.* If you want a front row seat, there is only one place to be.

Here.

Acknowledgments

THE AUTHORS WOULD LIKE TO THANK:

Dan Lazar at Writers House, for suggesting this most unlikely but ultimately fortuitous author-publisher *shidduch*.

Ilan, Valeria, Kezia, and the rest of the Gefen Publishing House team.

Meghan Stoll, Michelle Levy, Joel Weinberg, and Dan Lefkovitz for reading early versions of the manuscript and offering brilliant suggestions on how to make it better.

Rabbis Neal Katz and Brian Strauss for their support. It's always good to have a few rabbis in your corner.

The dozens of Israelis who allowed us to interview them for this book. It's one of the things we love most about Israel: that big-time athletes, artists, entrepreneurs, professors, journalists, and experts of every kind will actually sit down over coffee with a pair of authors they don't know and answer endless questions, simply because they believe in their mission.

*   *   *

Thank you to:

Benji Lovitt, for being such a wonderful collaborator.

My parents, for sending me to Jewish day school and summer camp, despite the expense. I have a clear memory of my father making out a check to Solomon Schechter Day School, circa 1983, and bemoaning the fact that a year's tuition at Schechter, about $4,000, was more than what he paid to go to college. Thank goodness the cost of Jewish education has since gone down.

For all they did to help make this book a reality: Miri and Ari Loren; Pam Feldstein and David Loewenberg; Caryn and Andrew Moss; Christine and Jeff Toback; Lynn Jablowsky; Ilanit and David Meckley; Bonnie and Manny Citron.

Zack Bodner, for helping us to better understand the reality of what we'll call "the financials." We so appreciate your guidance.

Jonathan Kessler, for the multiple conversations about Zionism, peoplehood, and other light topics. Your advice has been invaluable.

Yael Gargir and Mia Hancock, for all you do behind the scenes. How wonderful to have you in my corner.

The wonderful teachers and educators who strengthened my connection to Israel over the years: Charlotte Glass, Holly Rosenberg, Liat Citro, Tirtza Haviv, Batya Retsky, David Soloff, Pnina Mazor, and Ze'ev Harari.

My five favorite Israelis: Dorit, Stav, Noam, Yuval, and Yoni, for your love, your encouragement, and for never walking into my home office uninvited to tell me something important, like that we're out of parmesan cheese.

J.C.

* * *

Thank you to:

Joel Chasnoff, for your truly impressive discipline and work ethic.

My fellow *olim*, especially the ones who helped me along the way (like Talia Klein Perez, who met me and my stack of unpaid Hebrew bills at a café fifteen-ish years ago) and inspired me to pay it forward to the next generation. Nobody else knows what we go through. Being part of this family is a big reason I'm still here.

The sabras who've helped me and made this *oleh* feel less and less *chadash*.

All the *shlichim* I've met over the decades, especially through Young

Judaea. You represent the best of what this country has to offer. It's because of you that so many of us move here.

My Young Judaea Year Course gap year roommate, former boss, and close friend Neil Weidberg. When I revealed to him that I didn't enjoy living in New York, he asked me, "Why don't you go spend a year in Israel?" What in the world would I be doing now if you hadn't asked?

B.L.

Sabra Bingo!

NOW THAT YOU KNOW THE INS AND OUTS OF THE ISRAELI ethos, it's time to enjoy a game of Sabra Bingo!

Use the board on the following page, or print it at Israel201.com. Then check off the appropriate squares whenever you see these typical Israeli behaviors in action.

For even more fun, see if you can explain the root cause of each behavior. What is it about the Israeli psyche that makes them act this way?

S	A	B	R	A
Car parked on the sidewalk, possibly even blocking the path, requiring pedestrians to walk in the street	Someone asking you the value of your house/monthly rent, or whether your parents give you assistance	Naked baby at the beach	Store checkout clerk commenting on your purchases, asking what you plan to cook	Café/falafel shop owner offering complimentary food and drink to a soldier
Child asking you to help them cross the street	Someone appearing from nowhere, claiming to be ahead of you in line	Two people arguing about something seemingly inconsequential	Israeli male casually mentioning that he was in elite unit of the IDF	Bus driver letting a passenger on for free because they don't have enough money to pay
Two people haggling over a price	Total stranger giving you unsolicited advice	Someone wishing you "Shabbat Shalom" before Friday in anticipation of upcoming Sabbath	Man wearing jeans and sneakers to a wedding or bar/bat mitzvah	Stranger inviting you to a Shabbat/holiday meal
Someone smoking next to a "no smoking" sign, possibly indoors	Adult male swimming in his underwear at beach or public pool	Person asked for directions escorting you part or all of the way to your destination	Child age ten or under riding on the back of a motorcycle	Someone bringing food/drink from one establishment into another or to somewhere it doesn't belong
Motorcycle driving toward you on the sidewalk, possibly accelerating	Cabbie telling you why the current prime minister is incompetent	Car driving wrong way down one-way street in reverse	Child talking back to an adult other than their parent	Adult urinating in a public place

Words You Need to Know

THROUGHOUT *ISRAEL 201*, WE USE HEBREW WORDS, ALWAYS transliterated and defined upon first use.

Some words appear throughout the book and are critical in understanding Israel and the Israeli identity. They are defined here.

aliyah. The process of a Jewish person immigrating to Israel. The term *aliyah* literally means "ascent" or "going up," connoting that when a Jew moves to Israel, he or she is "going up" spiritually. Many immigrants have joked that not only do they go up in spirit, but in cost of living and level of stress.

Anglo. A catch-all for natives of English-speaking countries such as the United States, Canada, the United Kingdom, South Africa, and Australia. A little disconcerting for *olim* who are used to thinking of Anglos as White Anglo-Saxon Protestants.

Ashkenazi (plural: Ashkenazim). A Jew whose ancestors came from Central or Eastern Europe, including Poland, Russia, and Germany.

chutzpan **(male)/***chutzpanit* **(female).** One who exhibits chutzpah.

dati. Hebrew for "religious," but more specifically a Jew whom North Americans might call Modern Orthodox. Men cover their heads with a kippah (yarmulke), women may wear skirts and cover their hair, and all observe laws of Sabbath and kashrut.

gibbush. Generically: social cohesion. Specifically: the grueling IDF pre-induction test that future soldiers need to pass to secure a spot in Shayetet-13 (Israel's Navy SEALS), Paratroopers Brigade, Air Force Fighter Pilot Training, and other elite units.

Haredi (plural: Haredim). Ultra-Orthodox Jews, easily distinguishable by their dress: black hats, often long sidelocks, black suits and tzitzit (fringes) for men, and long skirts and head covering for women.

Knesset. Israel's Parliament.

meenus. Overdraft. From the English word *minus,* as in a negative bank balance.

Merkaz. Hebrew for "center," used by Israelis to describe the center of the country, what's sometimes referred to as "Hadera to Gadera." Including Greater Tel Aviv, the Merkaz is home to over two million people. This is important because Israelis sometimes speak about two Israels: the Merkaz, and everywhere else. A sad but true fact is that when Hamas targets the citizens of southern Israel with missiles, the government often chooses to retaliate only when the population centers of the Merkaz come under fire.

miluim. Reserve duty. Many Israelis go for regular stints of reserve duty in the IDF for up to several weeks annually.

Mizrachi (plural: Mizrachim). In Hebrew, *mizrach* means "east." Mizrachi describes Jews whose descendants came from lands east of Israel: Iraq, Iran, and Yemen.

oleh chadash (male)/*olah chadashah* (female)/*olim chadashim* (plural). A "new immigrant," one who has "gone up," i.e., made aliyah to Israel. Throughout the book, we often use the shortened version, meaning "immigrant": *oleh, olah,* and *olim.*

protektziah. Slang for "connections," or preferential treatment based on who you know. Also referred to as Vitamin P.

Rabbanut. The Israeli Rabbinate, the governmental body that rules over religious matters as well as some civil matters that they deem religious.

sabra. A native-born Jewish Israeli.

Sepharadi (plural: Sepharadim). A Jew whose ancestors were expelled from Spain and Portugal during the Inquisition of 1492 and moved to the Middle East and North Africa. Because many arrived and lived for hundreds of years in countries where Mizrachi Jews already resided, the terms *Sepharadi* and *Mizrachi* are often used interchangeably in Israeli society.

Shabbat. The Sabbath, the Jewish day of rest, which falls on Saturday.

soger et hachodesh. To finish the month, as in, to get to the end of the month with some money still left in your bank account.

And, finally, one more word in English:

Diaspora. The communities of Jews living outside of Israel, dating back to their exile following the destruction of the Second Temple in Jerusalem in 70 CE.

Additional Resources

CHECK THE BOOK'S WEBSITE, ISRAEL201.COM, WHERE WE'LL continue to upload video clips, educational guides, and other resources to complement your reading experience. It's also where you can contact us.

About the Authors

JOEL CHASNOFF is a stand-up comedian and author of multiple books, including the comedic memoir *The 188th Crybaby Brigade*, about his service in the Israel Defense Forces, and *Essential Tennis*, which he cowrote with coach Ian Westermann. To date, he's performed comedy at more than a thousand Jewish fundraisers and social events in ten countries. A Chicago native, Joel now lives in Ra'anana, Israel. Video clips, tour info, and more at www.joelchasnoff.com.

BENJI LOVITT grew up in Dallas, Texas, and immigrated to Israel in 2006. For the past fifteen years, he's performed stand-up comedy and delivered cultural presentations about Israel for audiences in North America, Israel, Europe, South Africa, and Australia. His unique comedic perspective on Israeli society and Israelis has led him to be an oft-featured guest on Israeli TV and radio and in media outlets including *USA Today*, BBC Radio, *Time Magazine*, and *The Atlantic*. More on Benji at www.benjilovitt.com.

Visit both authors at www.Israel201.com.